The United Kingdom
National Air Quality Strategy

Presented to Parliament
by The Secretary of State for the Environment, the Secretary of State for Wales
and the Secretary of State for Scotland by Command of Her Majesty

CM 3587   March 1997   £17.85

Contents

The United Kingdom National Air Quality Strategy

Part I

Chapter 1: Setting the Scene

Introduction
1. This Strategy marks a watershed in the history of measures to control and improve the quality of air in the United Kingdom. It builds upon two broad trends, which have come together to create the platform for the formulation of a more strategic and integrated approach to air quality issues. The first is the elaboration of the principles of sustainable development. In particular, the United Nations Conference on Environment and Development at Rio de Janeiro in 1992 ("the Earth Summit"), has led decision-makers to try to break down the barriers between environmental and developmental policy-making, and to develop strategic, objective-led means of managing change over the long term. The second is progress in recent years at national, European and international level, in our understanding of air pollution and in the development of new instruments to tackle it. This has come to a point where a more comprehensive framework for management of air quality is both necessary and possible.

Air quality and sustainability

2. Air quality is an issue of sustainability, as we strive to create an environment in which individuals and communities can thrive. Essential to that process is the continued improvement of the external conditions which affect human health. Agenda 21, the central text on sustainable development to come out of the Rio Earth Summit, dedicates one of its chapters to "Protecting and Promoting Human Health". The overall objective of that chapter is "to minimize hazards and maintain the environment to a degree that human health and safety is not impaired or endangered and yet encourage development to proceed". Air quality is identified as a key element in the reduction of health risks from environmental pollution and hazards.

3. There is an increasing understanding of what those risks to health are, and the kind of benefits to be gained from making the air cleaner. Fortunately the UK has moved on from the days when, as in 1952, air pollution could cause an estimated 4000 additional deaths of sick and mainly elderly people in just a few days - an effect comparable with that of a major influenza epidemic. Nevertheless, some recent statistical analyses suggest that, even at the substantially lower levels of airborne pollution we experience today, there are associations with premature mortality, chronic illness and discomfort for sensitive groups. On the other hand, there is no evidence that healthy individuals are likely to experience acute effects at typical UK air pollution levels. Steps to improve the quality of air will diminish any remaining risks, and provide a more pleasant living and working environment for us all.

4. This is not to say that air quality is solely an issue of human health; we know that air pollution can degrade both the natural and the man-made environment - forests, lakes, crops, wildlife, buildings and other materials can all suffer significant damage from high levels of airborne pollutants. Again, cleaner air will help to reduce the likelihood of any such damage and its economic costs. If we are genuinely "not to cheat on our children" our legacy to them must include acceptably clean air.

5. There are also close links between air quality issues and climate change. Some are areas of synergy, for example ozone is a greenhouse gas, so controlling it and its precursors can also contribute to our commitments to tackle climate change as can controls over emissions of nitrogen oxides from aircraft. However, there may also be areas of conflict,

for example where action to control vehicle emissions may reduce fuel efficiency. Sustainability requires the identification of policies which maximise the synergy and which strike a careful balance between the possible conflicts.

6. This was reflected in the UK's Sustainable Development Strategy. In acknowledging that good air quality was essential for human health and the well-being of the environment as a whole, it identified one of the key issues for sustainability as "to manage local air quality, especially in urban areas, and in particular to ensure that all relevant sectors - industry, transport, local authorities and the general public - contribute." All of these, as well as central government and the new Environment Agencies, have a part to play to secure a sustained improvement in air quality which in turn can bring about a lasting improvement in the quality of life.

Developments in air quality understanding and instruments

7. Meanwhile, the conditions for strategic, objective-led management of air quality are coming into place. More extensive monitoring data and information on air pollution in the UK is becoming available. This is being matched by greater scientific understanding of its origins and effects, and increasing attention to the economic analysis of the associated costs and benefits. This progress has coincided with, and informed, the development of a wider range of instruments intended to manage air pollution. Each of the main sectors, which are sources of emissions, has been brought within a regulatory framework. Substantial reductions in emissions of a wide range of pollutants have already been achieved:

- the task of controlling emissions from domestic sources is now well towards completion;

- a framework for managing and reducing industrial emissions is now in operation under the Environmental Protection Act 1990 (EPA 90), operated by both the new Environment Agencies and local authorities; and

- policies adopted by the Government in the last few years have marked the development of a more comprehensive approach to the control and improvement of vehicle emissions.

8. At a European level the history of air pollution control measures, which stretches back more than twenty years, is now being consolidated into a broad European framework for the management of air pollution for the next decade and beyond. Over the last decade, a framework of international treaties has been developed to cover most of the transboundary pollutants that affect the UK.

9. Central to the further development of air quality policy, both in this Strategy and in other fora, is an understanding of the relationship between the different levels at which air pollution is generated and, in consequence, controlled. Many of the pollution control strategies developed in the late nineteen seventies and eighties were primarily directed at tackling long-range transboundary pollution, and acid rain in particular. As described in greater detail in the following chapter, this meant that instruments were developed to meet national emission reduction targets. However, emerging evidence that there remained problems associated with personal exposure to ambient concentrations of air pollution led to a new interest in local air quality and local sources of pollution.

10. The relationship between transboundary and local air pollution is difficult to specify. For those pollutants, such as sulphur dioxide (SO_2) and nitrogen oxides (NO_x), which were initially controlled for their transboundary effects, the abatement of emissions from major sources has contributed to the achievement of improved local air quality. The contribution of transboundary sources to local pollution is, for these and other pollutants such as particles, potentially significant. In the case of ozone, the contribution of transboundary sources to ambient levels is so great as to necessitate internationally coordinated action for the control of its precursors. With the continued abatement, however, of these sources, the management of local air quality needs also to look at the contribution made by local sources, which is in many cases dominant, particularly during episodes of elevated pollution levels.

The Air Quality Strategy

11. While great progress has been made, and continues to be made, in the improvement of air quality, important long and medium term goals remain, which will lead to significant, further reductions in the number and extent of episodes of poor air quality, both in summer and in winter. These goals are to be achieved with due regard to the need to balance, as far as knowledge allows, any costs and the ensuing benefits.

12. Many of these issues have already been foreshadowed in the Government discussion documents *Improving Air Quality* and *Air Quality: Meeting the Challenge.* In the light of the debate stimulated by those discussion documents, and in the light of the developments described above, the Government concluded that it should formulate, within a general strategy, its standards, objectives and targets for the improvement of air quality, the main policies which currently reflect them, and the process by which, over reasonable planning horizons, the Government aims to move towards those targets. Accordingly, among the provisions included in the Environment Bill was the management and improvement of air quality in the United Kingdom. The Bill received Royal Assent in 1995.

13. The Environment Act 1995 also laid the foundations for a nationwide system of local air quality management, in which local authorities are obliged to review and assess the quality of air in their areas, and to take action where air quality standards or objectives are breached or at risk of being breached. Such standards or objectives are to be defined by means of regulation.

Structure and Scope of the Strategy

14. This document provides the Strategy required under that legislation. It is set firmly within the established UK approach to air quality policy, following an effects-based approach. That approach is founded on the scientific assessment of the impacts of pollution; the derivation of the standards which embody a high degree of protection of human health, and implementation through proportionate, targeted action which weighs the expected benefits against the associated costs. The fundamental principles underlying the Government's approach to air quality policy are set out in Box 1.A.

15. Part II of the Strategy contains, for eight priority pollutants: a summary of the assessment undertaken by the Government's medical and scientific advisers of the standard appropriate for the protection of health; an assessment of the current prevalence of the pollutant within ambient air;

and the improvement which can be expected from policies and technologies in place or planned; and, where necessary, a review of what more needs to be done to secure the standard and the practicality of doing it. These assessments provide the basis, in Part I Chapters 2-7, for elaboration of the Government's Air Quality Strategy: its objectives; the standards on which they are based; the respective roles of central and local government; and the contribution that can be made by the industrial and transport sectors of the economy and by the community.

16. Following these principles, and in line with the framework devised in the Environment Act 1995, the Strategy is tightly targeted on the management of ambient air quality. Therefore, it is not designed to encompass issues which are related - other indicators of environmental quality such as water, soil or noise pollution, for example, or environmental problems to which emissions to air may contribute, such as the deposition of airborne pollutants or eutrophication. The chapter on the international context for the Strategy does, however, give some coverage to policies on acidification.

17. Neither occupational exposure nor indoor air quality are included in the scope of this Strategy. The total personal exposure of an individual to an air pollutant may be significantly influenced by indoor exposure. A substantial body of research to investigate exposure to air pollutants in the indoor, non-occupational environment is being supported by the Department of the Environment.

18. Whilst this Strategy has been drafted in United Kingdom terms and outlines a unified approach, there will be circumstances where the different arrangements applicable in Northern Ireland may require that a different approach is adopted. For example, Northern Ireland has its own environmental legislative code and the Environment Act 1995, which sets out the legislative structure for the Strategy, does not apply. Corresponding legislation will be prepared for Northern Ireland under the Order in Council procedure.

Conclusion   19. The aim of this Strategy is to map out, as far as possible, the future of ambient air quality policy in the United Kingdom at least until the year 2005. A particular purpose is to ensure that all those who contribute to air pollution, or are affected by it, or have a part to play in its abatement, can identify both what is statutorily required from them and what further contribution they can voluntarily make in as efficient a manner as possible. Vital to this process is the notion that the Strategy must be evolutionary rather than a rigid structure determined by the conditions pertaining at the present time. The Act requires that the Secretary of State's policies on air quality are regularly reviewed, and it is the Government's intention to initiate the first review of this Strategy in 1999. Preparation for this first review will be assisted by the establishment of an Air Quality Forum, which will bring together representatives of all interests to ensure that the implementation of current policies is carefully monitored and reviewed, and that future priorities can be identified.

Chapter 1: Setting the Scene

Box 1.A: **Principles of Air** **Quality Policy**	The UK Government believes that air quality policy should in general be based on the following principles:
Sustainability	It is a fundamental precept that policy should seek to drive technologies, behaviour and use of resources towards modes of operation which are sustainable in the long term.
Effects-based approach	The touchstone for action should be environmental objectives, expressed in terms of environmental quality. This allows areas to be treated proportionately to their particular risk of damage using the package of measures most suitable for them to achieve the agreed objectives. Effects include those on human health as well as on the natural and man-made environments.
Risk assessment	Quality objectives must be set on an understanding of the relationship between exposure to levels of pollution and their effects. This enables judgement to be made on critical loads and critical levels[1] and can inform decisions where there are no critical thresholds (i.e. where effects occur at all loads or levels) or where the costs of meeting critical levels are higher than the benefits.
Sound science	Risk assessment must be based on internationally robust scientific evidence, published and peer reviewed.
Proportionality	Where the case for action is adequately made, the measures concerned should be proportionate to their objectives, in the light of an assessment of the costs and benefits involved. Measures should provide for flexibility in implementing and enforcing international obligations.
Polluter pays principle	The cost of measures decided by authorities to ensure that the environment is in an acceptable state should be reflected in the cost of goods and services which cause pollution in production and/or consumption.
Precautionary principle	Where there are significant risks of damage to the environment, the Government will be prepared to take precautionary action to limit the use of potentially dangerous materials or the spread of potentially dangerous pollutants, even where scientific knowledge is not conclusive, if the balance of likely costs and benefits justifies it.
INTERNATIONAL	
Subsidiarity	Action should only be taken at EC level where a Community objective cannot sufficiently be achieved by member states, taking account in particular of transnational aspects.
Effective International Monitoring and Enforcement	Ratification of UNECE Protocols should be followed by national action plans, where appropriate, and reports to the relevant supervisory body. The implementation of EC legislation should be rigorously monitored and enforced across the Community.

[1] The Protocol to the 1979 UNECE (UN Economic Council for Europe) Convention on Long-Range Transboundary Air Pollution on the Further Reduction of Sulphur Emissions, 1994 gives the following definitions:
Critical loads - a quantitative estimate of exposure to one or more pollutants below which significant harmful effects on sensitive elements of the environment do not occur, according to present knowledge;
Critical levels - the concentrations of pollutants in the atmosphere above which direct adverse effects on receptors such as plants, ecosystems or materials may occur, according to present knowledge.

Chapter 2: The International Context

Introduction 1. Any strategy for air quality in the UK must take account of the international context in which air quality and pollution policies are set. The UK is subject to international responsibilities and rights in this field both because of the simple fact that the passage of air does not confine itself within national boundaries, and through our membership of international institutions, particularly the European Union and the United Nations Economic Commission for Europe (UNECE).

2. In general, the long-range movement of air masses across the earth's surface means that pollution emitted in one country is, to a significant extent, shared with its neighbours. For some pollutants, such as sulphur dioxide (SO_2) or nitrogen oxides (NO_x), emissions can travel much further than a country's immediate neighbours. The winter pattern for Northern Europe is for air masses to travel eastwards from the Atlantic, over the British Isles towards France, Germany, Benelux and Scandinavia. The UK therefore "exports" a significant proportion of some of the pollutants emitted in this country. Recent estimates suggest that more than 75% of the UK's SO_2 emissions and as much as 90% of our NO_x emissions are transboundary, leading to deposition either in other countries or in the sea.[1] These pollutants are often converted through atmospheric chemical reactions into secondary pollutants, such as sulphates and nitrates which are the primary cause of acidification.

3. This is not the only pattern of air pollution movement. There are flows in the other directions, including northwards and westwards from mainland Europe over Britain. Around 40% of the deposition (rather than ambient levels) of oxidised nitrogen in the UK originates from sources outside the UK,[2] and up to one half or more of peak ambient ozone (O_3) levels in Southern Britain originate outside the UK. It is not only the traditional long-range pollutants which are transboundary. In Europe generally, there are cross-border flows of pollution wherever sources are close enough to national boundaries; Britain has an exchange of more 'local' pollutants with other countries, especially northern France and Benelux.

4. The consequence of these complex inter-relationships is that, for strategies to reduce levels of transboundary pollutants to be effective, there must be a sufficient degree of international cooperation. Policies applied nationally to reduce the ambient levels or deposition load of some pollutants can be rendered fruitless unless there are equivalent efforts to control "imported" pollution. For a strategic approach to controlling long-range transboundary pollution, such as acid rain or ozone, international cooperation is the only option. Furthermore, action to abate emissions of many pollutants may impinge upon other international obligations.

5. It also follows that the development of national strategies for the control of polluting emissions and the management of ambient air quality must reflect the constraints and opportunities which measures and agreements at international level provide. International agreements to reduce national emissions of transboundary pollutants, for example, clearly provide the basis for control strategies at national and local level.

---

[1] Source: EMEP (Cooperative programme for monitoring and evaluation of the long-range transmission of air pollutants in Europe); Ten years calculated fields and budgets, July 1995. Figures are for 1993. The proportions of total UK emissions deposited in the sea are calculated as: SO_2 -41% and NO_x - 35%.

[2] Source: as for 1

In some cases, such as European Community (EC) vehicle emission standards, internationally agreed measures can be an important instrument for the achievement of national objectives. The National Strategy will take account of the international framework as it evolves - there is an increasing movement towards the globalization of air quality standards and of control technologies, harmonisation of the methods of measuring air quality and emissions and other management tools, driven both by legislation and by the work of bodies such as the European Environment Agency and its Topic Centre on Air Quality. This trend cannot be ignored, even where the changes are not enforced through EC legislation.

6. The UK plays an active role in both principal international fora in which air quality strategies and policies are discussed and determined, the UNECE and the EC. This Strategy takes full account of the commitments and responsibilities of these bodies, and in relevant areas looks to further action by them to help achieve its objectives.

7. The UNECE covers the whole of the European region from the Atlantic coast to the former Soviet Union, the Balkan states and Turkey. The USA and Canada are also members. Of its fifty-four Member States, forty, plus the EC, are Parties to the 1979 Convention on Long-Range Transboundary Air Pollution (UNECE/LRTAP Convention) which entered into force in 1983. Under this Convention, which lays down general principles for international cooperation on the abatement of air pollution, the Parties have adopted a number of Protocols designed to achieve emission reductions for specific pollutants. Details of the Protocols already adopted are given in Box 2.A. As can be seen, the approach usually employed in

BOX 2.A:
UNECE Convention on
Long-Range
Transboundary Air
Pollution

The Convention on Long-range Transboundary Air Pollution was adopted on 13 November 1979 and entered into force on 16 March 1983. It has been followed by a series of protocols that have laid down more specific commitments for parties.

■ The *Protocol on Long-term Financing of the Cooperative Programme for Monitoring and Evaluation of Long-range Transmission of Air Pollutants in Europe (EMEP)* was adopted on 28 September 1984 (UK ratified on 28 August 1985) and entered into force on 28 January 1988. It commits parties to mandatory annual contributions to the EMEP budget approved by the Convention's Executive Body.

■ The *Protocol on the Reduction of Sulphur Emissions or their Transboundary Fluxes by at least 30%* (Helsinki Protocol or "30% club") was adopted on 8 July 1985 and entered into force on 2 September 1987. It commits parties to a 30% cut in total national SO_2 emissions by 1993, based on 1980 levels. Although it did not become a Party, the UK achieved a reduction of 37% by the end of 1993.

■ The *Protocol concerning the Control of Emissions of Nitrogen Oxides or Their Transboundary Fluxes* (Sofia Protocol) was adopted on 31 October 1988 (UK ratified on 15 October 1990) and entered into force on 14 February 1991. It commits parties to bringing NO_x emissions back to their 1987 levels by 1994. The UK has met this target.

■ The *Protocol concerning the Control of Emissions of Volatile Organic Compounds (VOCs) or their Transboundary Fluxes* was adopted on 18 November 1991 (UK ratified on 14 June 1994) but it has not yet entered into force. It commits most parties to secure a 30% reduction in VOC emissions from 1988 levels by 1999.

■ The *Second Protocol on the Further Reduction of Sulphur Emissions* was adopted in Oslo on 14 June 1994 (UK ratified on 13 December 1996) but has not yet entered into force. The protocol requires different percentage reductions from parties depending in part on the quantity of their emissions, but importantly takes into account the nature of the impact upon the environment. This is based upon the concept of critical loads and requires the UK to make reductions in sulphur emissions, against 1980 levels, of 80% by 2010.

these Protocols is a set of agreed national emission reduction targets. In the Second Sulphur Protocol, these targets were based on the concept of critical loads; it is planned that this concept, which underpins the effect-based approach, will also be used in the negotiation of a new multi-pollutant, multi-effect Protocol.

8. The EC has been legislating to control emissions of air pollutants and to establish environmental quality objectives for the last two decades. There is presently a wide range of instruments related to air quality and pollution control. These cover environmental quality standards, vehicle emission standards, fuel quality standards, industrial pollution control and environmental impact assessment.

9. Most recently, the EC Council of Ministers has adopted two directives central to European air and pollution policy:

■ the Ambient Air Quality Assessment and Management Directive establishes a framework under which the Community will agree air quality limit or guide values for specified pollutants in a series of "daughter directives". These will supersede existing air quality legislation. Under this "framework directive", Member States will have to monitor levels of these pollutants, and draw up and implement reduction plans for those areas in which the limit values are being or are likely to be breached. It is envisaged that the structures established under the Environment Act 1995 and the National Air Quality Strategy will provide the principal means of carrying out the UK's commitments under this directive. The first daughter directives are due to be proposed in the summer of 1997; and

■ the Integrated Pollution Prevention and Control Directive will be the cornerstone of European industrial pollution control policy. It is a source-based Directive, requiring Member States to ensure that major industrial installations receive permits based on the Best Available Techniques for pollution control, subject to technical and economic feasibility, taking into account discharges to all environmental media. The UK already has a system of Integrated Pollution Control, introduced in the Environmental Protection Act 1990 which, modified accordingly, will form the basis of our implementation of the Directive. The Directive came into force on 30 October 1996 and must be transposed by Member States no later than 30 October 1999.

The UK's approach to international commitments and priorities for the future

10. The Government believes that the effects-based, flexible approach taken within the UNECE provides a rational framework for tackling the problems of transboundary air pollution. Target emission reductions for each country are derived from agreed environmental objectives which are derived from an assessment of the costs and benefits of varying levels of action. It is then left to each Party to devise its own strategy for achieving the reduction to which it is committed. This approach has a number of advantages. First, it provides the flexibility to allow for a full range of policy instruments to be applied in the most cost-effective way. Second, it is consistent with the principles of subsidiarity. Third, it minimises the administrative burden of regulation at the international level. However, for this approach to work, there must be confidence in Parties' willingness and ability to meet the targets set.

11. There are a number of important developments in international air quality policy in which the UK will need to take an active role. The

negotiation of the UNECE/LRTAP Convention's multi-pollutant, multi-effect Protocol, due to be completed during 1998, will be of central importance because it will attempt, in developing targets for emission reductions, to take into account the several effects of NO_x, ammonia and volatile organic compound (VOCs) emissions. One of these effects is the contribution NO_x emissions make to the problem of acidification or acid rain.

12. Acid rain was one of the earliest air pollution problems to be internationally recognised and tackled, as a long-range transboundary issue. However, emissions of SO_2 and NO_x in particular are still contributing to damage to both the natural and man-made environments through acidification. Emissions in the UK and across the rest of Europe have fallen dramatically since the 1970s, particularly in the light of the first UNECE Sulphur and NO_x protocols and work within the EC to curb emissions from large combustion plants, from other industry and from vehicles. Yet critical loads - the measure of how much acidifying deposition different types of environment can tolerate - are still being exceeded across the European region, including parts of the UK. Although it is impossible, given background levels, to eliminate exceedences of critical loads as currently defined, it is generally accepted that more needs to be done. The UK was among the first countries to ratify the Second Sulphur Protocol which commits us to a cut of 80% in sulphur dioxide emissions on 1980 levels by the year 2010 and, as stated above, the forthcoming multi-pollutant, multi-effect Protocol will also be a crucial instrument. At EU level, the UK has supported the principle of developing a coherent framework for tackling acidification. The European Commission is currently developing such a strategy. We are encouraging it to ensure that it is practical, achievable and scientifically robust. We hope to ensure that forthcoming EC legislation on emissions from large combustion plants, on the sulphur content of liquid fuels and on vehicle emissions will be developed where necessary with due regard to the promulgation of a wider, integrated and cost-effective acidification strategy.

13. The other main issue dependent on the outcome of the multi-pollutant, multi-effect Protocol is ground-level ozone, the main component of summer smog. As mentioned in the introduction to this section, ozone also has an important transboundary dimension. Ozone is also a greenhouse gas and so action to control its production may also contribute to meeting the UK's climate change commitments. It is a secondary pollutant which is caused by chemical reactions, triggered by sunlight, involving other "precursor" pollutants. Emissions of its precursors, the principal ones being NO_x and VOCs, contribute to ozone levels up to one thousand or more kilometres away from their source. Levels of ozone have tended to increase in recent years, with a number of ozone "episodes" each summer in most European Community countries.

14. The number of sources of ozone precursors is so great, and their distribution so diffuse and widespread, that only action coordinated on a grand scale will be sufficient to tackle the problem effectively. This is increasingly recognised, for example at the recent Ministerial Conference on the subject held in London, which involved eight countries of northwest Europe,[3] the UNECE, the European Commission and the European Environment Agency. This meeting developed several joint steps on

---

[3]   Belgium, Denmark, France, Germany, Ireland, Luxembourg, the Netherlands and the UK.

technical issues, set a target for ozone reduction and called on the UNECE and the EC to move faster towards a pan-European ozone strategy, something for which the UK has been pressing in the international sphere.

15. The first priority for the UK is to secure a sensible and constructive agreement on emission reductions across Europe in the new multi-pollutant, multi-effect Protocol. In order to ensure that action to combat ozone formation is properly coordinated, the UK has supported the extension to cover emissions of VOCs. The reductions should be based upon the latest available verified understanding of critical levels for ozone and appropriate cost-benefit analysis.

Forthcoming EC proposals

16. The other major area in which the Government is and will be taking a major role is in the development of daughter directives under the EC framework Directive. The Commission established working groups, consisting of Member States, industry, non-governmental organisations and other interested institutions, to prepare recommendations for directives for the first group of pollutants in the framework Directive's list. The UK chaired the group on particles, jointly with Germany, and was also a member of the group working on SO_2. The Government provided all the working groups with information on the UK's work on air quality standards and assessment and management; in particular, it commissioned work to identify and examine cost-effective instruments to deliver possible new EC air quality limits for SO_2. The first proposals are due in the summer of 1997.

17. There will be a number of additional proposed directives on air quality issues for discussion in the EC in the next year or so. These include new vehicle emission and motor fuel quality standards arising from the European Commission's Auto-Oil programme and the control of industrial emissions of VOCs from solvent use. Further proposals may include a revision of the Large Combustion Plants Directive, and a lower limit on the sulphur content of liquid fuels. In so far as they are not taken into account in this Strategy, the outcome of the negotiations on these proposals will be taken into account in its first review. For those proposals aimed primarily at the tackling of transboundary air pollution, the Government will seek to ensure that they form part of a consistent and cost-effective strategy.

18. Of these proposals perhaps the most significant for ambient air quality are those on vehicle emission and fuel quality standards for the year 2000 and 2005, arising out of the Auto-Oil programme. This programme saw the cooperation of the Commission with the European oil and automotive industries in assessing which measures to reduce traffic pollution would be the most cost-effective in achieving air quality objectives, derived from World Health Organisation guidelines and other expert recommendations, by the year 2010. The aim was to provide a scientific and cost-effective basis for proposing new, tighter EC vehicle emission and fuel quality standards. These proposals will also have significant implications for transboundary pollution (notably acidification and ozone formation), as the principal pollutant targeted by the programme was nitrogen dioxide (NO_2). The UK Government broadly supports the European Commission's first two proposals, on passenger cars and on fuel quality, and our position reflects our assessment both of the need to meet the air quality objectives set out in this Strategy and the technical and economic feasibility of the

standards proposed. Further directives on light commercial vehicles, heavy duty vehicles and inspection and maintenance are due later in 1997. Action to reduce vehicle emissions can sometimes lead to reduced fuel efficiency and hence higher emissions of carbon dioxide, the most important greenhouse gas. At the same time, EC proposals also include a strategy to improve significantly the fuel efficiency of new cars which should not be achieved at the expense of local air quality. This is an area where a careful balance needs to be struck between local and global environmental priorities.

19. The European Commission has recognised the importance of providing an assessment of the costs and benefits of its proposals. This is an important and positive step. As stated in the first Chapter, the Government believes that international air pollution policy should be judged according to the set of principles set out in Box 1.A. It will seek to ensure that measures agreed in the EC are consistent with these principles in order to help develop European legislation which provides an efficient means of achieving European environmental objectives.

Conclusion 20. An air quality strategy which is devised at a national level is inextricably enmeshed with the international framework in which it is set. The points of particular salience are:

■ transboundary air flows mean that objectives for certain pollutants - notably ozone - may themselves only ultimately be achievable through international co-operation;

■ existing goal-based international strategies, and the national plans for their implementation, provide a useful precedent for the development of this fuller strategy;

■ the measures and timescales developed within those strategies for implementing our commitments on transboundary deposition can significantly contribute to achieving air quality objectives in the UK; in some cases, European legislation on pollution abatement will be the principal means of achieving UK objectives;

■ in turn, the UK's National Strategy will need to accommodate the measures required to deliver its international commitments;

■ as emphasised, national and international strategies will increasingly have to converge, particularly in the development of effects-based strategies under the UNECE, and in the application of the EC Directive on Ambient Air Quality Assessment and Management.

Chapter 3: Setting Standards and Objectives

Introduction

1. For many years the ultimate aim of air quality policy in the United Kingdom has been to ensure that polluting emissions, and ambient air quality generally throughout the country, do not cause significant harm to human health and the environment (see Box 3.A). Air quality in the UK is, in general, already very good. This Strategy identifies and addresses the remaining problems.

Box 3.A: Rendering Emissions Harmless

The "rendering harmless" approach can be traced back to the first use in national pollution legislation of the Best Practicable Means (BPM) concept, in the Alkali Act 1874. The obligation, now in the Health and Safety at Work etc Act 1974, is on process operators to use BPM for preventing emission of noxious or offensive substances "and for rendering harmless and inoffensive such substances as may be so emitted". The same principle is now enshrined in the Environmental Protection Act 1990. In this, the requirement is on the regulator to impose conditions aimed at securing use of the Best Available Techniques Not Entailing Excessive Cost (BATNEEC) for preventing and reducing releases of specified substances and "for rendering harmless such substances which are so released".

2. Establishment of a general air quality strategy which attempts to map out the path to achieving that aim requires a simple and clear conceptual framework which is readily accessible to all the parties on whom the Strategy bears. The fragmented development of European and UK legislation and policy has seen the introduction of a wide range of concepts, whose application and relation to each other is not clear or readily understood. They include standards, objectives, targets, limit values, guide values and various others. While a number of these concepts will be employed, with their customary meaning or special definition, two key concepts provide the central structure for this Strategy - standards and objectives.

Standards and Objectives

3. The Environment Act 1995 requires this Strategy to include statements on "standards relating to the quality of air", and "objectives for the restriction of the levels at which particular substances are present in the air". Standards have been used as benchmarks or reference points for the setting of objectives. Standards are defined for the purposes of this Strategy as follows:

- standards are the concentrations of pollutants in the atmosphere which can broadly be taken to achieve a certain level of environmental quality. The standards relating to the quality of air are based on the assessment of the effects of each pollutant on public health.

4. Given a set of air quality standards as defined above, the Government must decide how they should inform air quality policy. Objectives are defined for the purposes of the Strategy as follows:

- the objectives provide policy targets by outlining what the Government intends should be achieved in the light of the air quality standards.

5. The objectives are necessary for the implementation of Part IV of the Environment Act 1995, and also are in line with the commitments the Government has made on targets in the 1996 and previous Environment White Papers. Objectives are generally expressed as a given ambient concentration to be achieved within a specified timescale.

6. It is the definition of these objectives which drives air quality policy and, in particular, the implementation of Part IV of the Environment Act 1995. First, the Government's air quality policies in general, including its negotiating position in European and international discussions, will be directed at the achievement of these objectives. Second, according to the terms of the Environment Act 1995, the Environment Agency, Scottish Environment Protection Agency (SEPA) and local authorities are required to have regard to this Strategy when exercising their pollution control functions under the Environmental Protection Act 1990.

7. However, the inclusion of standards and objectives in the Strategy does not in itself impose obligations on local authorities regarding the designation and operation of Local Air Quality Management Areas (AQMAs). This will happen only when there are standards or objectives prescribed by regulation under section 87 of the Environment Act. The framework introduced by the Environment Act 1995 envisages that, where the objectives are not likely to be met through national action, there should be complementary action at a local level. Therefore the Government has decided that the objectives should, apart from ozone, be included in regulations as the objectives which would trigger the designation of AQMAs as specified by section 83 of the Environment Act 1995. Under the Act the relevant local authorities would then be required to draw up action plans, for exercise in pursuit of the relevant objectives.

The Setting of Standards and Objectives

Standards

8. From the discussion in the previous section it follows that standards, as the benchmarks for setting objectives, are set purely with regard to scientific and medical evidence on the effects of the particular pollutant on health, or in the appropriate context, on the wider environment, as minimum or zero risk levels. Costs and benefits, and matters of current technical feasibility come into play at the later stage - in setting objectives and timescales, where they are an essential consideration. In the area of effects on human health this is the approach adopted by the World Health Organisation (WHO) in their formulation of their air quality guidelines published in 1987 and their subsequent revision in 1994/95, and by the Expert Panel on Air Quality Standards (EPAQS) in the UK.

9. The Expert Panel on Air Quality Standards was set up in 1991, fulfilling a commitment made in the Environment White Paper of 1990, "This Common Inheritance". The Panel consists of independent experts appointed, taking account of the advice of the Chairman, for their medical and scientific expertise. In performing its role as advisors to the Government on air quality standards[1], EPAQS has reviewed the published and peer reviewed evidence available, in order to provide the Government with recommendations for air quality standards.

---

[1] The terms of reference for EPAQS are: "To advise, as required, on the establishment and application of air quality standards in the United Kingdom, for purposes of developing policy on air pollution control and increasing public knowledge and understanding of air quality, taking account of the best available evidence of the effects of air pollution on human health and the wider environment, and of the progressive development of the air quality monitoring network".

10. The setting of air quality standards cannot be an absolutely precise exercise. Elements of uncertainty and judgement are unavoidable. Nevertheless a standard - in the sense of a defined level which can be taken to avoid significant risks to health - remains essential, and a necessary basis for meeting in practical terms, the statutory requirement to render pollution harmless. Clearly, however, such standards must be based on the best available scientific understanding and experience.

11. For some pollutants, it is possible to identify a concentration at or below which effects are unlikely even in sensitive population groups. At the present state of knowledge, this has been the case for ozone (O_3), sulphur dioxide (SO_2), carbon monoxide (CO) and nitrogen dioxide (NO_2). EPAQS has therefore recommended such concentrations as the standard for those pollutants. In other cases it is not possible to identify levels at which there is zero risk. This is true for genotoxic carcinogens[2], such as benzene and 1,3-butadiene, and so far, for particles. In recommending standards for benzene and 1,3-butadiene, therefore, EPAQS has assessed the published health effect evidence and has attempted to derive, using a widely accepted toxicological approach, a level at which the risk to public health would be exceedingly small.

12. The current state of understanding of the adverse health effects of particles suggests that, while associations between particle concentrations and a range of health effects have been demonstrated, it has not up to now been possible to derive a clear cut-off point below which there are no effects. Faced with this, the WHO, in revising their air quality guidelines, felt unable to recommend a single value for a guideline, but rather presented a linear exposure-response relationship based on published epidemiological studies. In recommending a particles standard for the UK, EPAQS supplemented the published studies used by the WHO with similar studies in the UK and derived a concentration of particles (based on PM_{10} - particulate matter with a mean diameter of 10 μm or less) which, in its judgement, would minimise the risk of health impacts on the population.

13. This approach by EPAQS towards pollutants for which no zero risk level is readily identifiable as a practical and sensible one, given the need for clear standards and objectives in the development of this Strategy, and given the requirement for the Government to move towards a practical realisation of its general criterion of harmlessness for its pollution policy. EPAQS represents the best available consensual medical advice in this field. Where EPAQS recommendations for a specific pollutant differ from WHO proposals, full discussion of this is given in the respective Chapter in Part II of the Strategy.

■ **The Government is therefore using the EPAQS recommendations, where they exist, as the air quality standards on which the setting of objectives will be based. Where EPAQS has not made a recommendation, the relevant information from WHO has been used, where available.**

Issues of Interpretation

14. In considering standards on a pollutant by pollutant basis, it is important to bear in mind the possibility of additive and even synergistic effects. There have been relatively few studies of such effects but the

2 Genotoxic carcinogens: substances which can cause cancers by attacking the genetic material.

Department of Health Advisory Group on the Medical Aspects of Air Pollution Episodes has recently reviewed the subject and concluded that there is no evidence for synergistic effects of mixtures of air pollutants, and that such evidence as is available suggests that the effects of mixtures are additive. This conclusion therefore lends support to consideration of standards and objectives on a pollutant by pollutant basis.

Objectives

15. Although it is the Government's ultimate objective to "render polluting emissions harmless", there needs to be a clearer statement of the Government's intentions for air policy in the medium term, or period over which this Strategy should be applied. As discussed above, the objectives of air quality policy need to be framed on the basis of the standards adopted, with due regard to consideration of the costs and benefits, and the feasibility and practicality of moving towards those standards.

16. The Government's primary objective is to ensure that all citizens should have access to public places without risk to their health and quality of life, where this is economically and technically feasible. EPAQS has had regard to this approach in recommending air quality standards and has taken account of possible effects on sensitive groups. The implication of this objective is that air quality policy should be directed towards getting air quality as close to the benchmark standards proposed here as is reasonable and justifiable on consideration of the costs and benefits, where those standards are not already being met.

17. First, a timescale needs to be set over which the Strategy is to be applied, and within which the objectives are intended to be achieved. **The Government has concluded that the end date for the current air quality Strategy should be 2005, as foreshadowed in** *Air Quality: Meeting the Challenge*. It considered whether the date set for the specific objectives might be earlier. However, if many of the objectives are to reflect health-based standards, substantial improvements in vehicle emissions will be required which can only be agreed in the European Community, and on which formal negotiations are not yet concluded. Technology and industrial lead times mean that the improvements the Government supports could not now be implemented and have effect much before the 2005 deadline. Nonetheless, wherever it is practicable and cost-effective, either through national policy or at more local level, improvements in pollutant concentrations which might secure achievement of the objectives before 2005 should be carried out. The aim should be to have a steady decrease in ambient levels of pollutants towards the objectives over the period of implementation. Where objectives are secured before 2005 the aim must be to sustain them.

18. In Part II of this Strategy document, there is a detailed analysis of each pollutant for which the Government is establishing air quality standards and objectives. In each case this analysis provides:

- a description of the pollutant and its effects;

- the sources of the pollutant;

- recommended standards and guidelines;

- current levels and predicted trends in the light of current policies; and

- a conclusion on the objectives which are to be adopted, where those objectives should apply and the feasibility of meeting them.

19. The Government is therefore adopting objectives which present a quantified assessment of the quality of air which it is intended that policies should be developed to achieve by the year 2005. It should however be made clear that in a number of cases, outlined in Chapter 4 below, considerable uncertainties remain as to whether these objectives are achievable. Although the objectives will be met in much of the country, at present it is not clear that future European legislation combined with other measures will be sufficient to achieve the objectives proposed everywhere.

20. For nitrogen dioxide (NO_2) and PM_{10}, for example, present estimates point to a gap in some areas of up to 10% between the reductions required to meet the objectives and those likely to be achieved by measures so far agreed to meet them. However, the size of this gap could be subject to significant variation depending on factors such as the weather, the accuracy of central forecasts of the emissions reductions to be achieved from planned measures; and the extent of emission reductions that may be achievable through reasonable local actions such as tougher enforcement of emissions standards and traffic management measures, which are difficult to quantify at this stage. These uncertainties are compounded by the fact that other relevant factors are hard to predict with certainty. Such factors include future patterns of consumer behaviour (particularly driving patterns), advances in scientific or medical understanding of the effects of air pollution, and the outcome of further analysis of the costs and benefits associated with the measures needed to achieve the objectives.

21. Although the objectives set out in Table 3.1 represent the Government's present judgement of air quality targets which are generally achievable, having regard to available evidence on costs and benefits, because of the uncertainties referred to above, a number of objectives proposed below should, in certain particular respects, be regarded as provisional. By this is meant that they are more likely than the other objectives to be changed by a future modification of this Strategy. Unless and until they are changed they are to be treated in just the same way as the other objectives. But they will receive special attention in the general review of this Strategy in 1999, particularly in relation to the costs and benefits of alternative measures and their relationship to the objectives, so that account can be taken of any developments which may help to resolve the current uncertainties. A good deal of work is currently in progress or planned on the costs and benefits associated with measures needed to achieve the objectives. If relevant evidence becomes available before the 1999 review, the Government will consider how far it is sufficiently significant to justify, exceptionally, reconsideration of the objectives ahead of the review.

Application of the objectives

22. The Government considers that **the objectives should apply in non-occupational near-ground level outdoor locations where a person might reasonably be expected to be exposed over the relevant averaging period.** Policies which aim at achieving air quality objectives at points where the highest measurable concentrations prevail with no regard for whether or not a person might be exposed would be inappropriate and highly inefficient. Bearing these considerations in mind, assessing where objectives are likely to be achieved is likely to be a major task. While central Government can go a long way towards making such assessments there will inevitably be a limit to the detail at the local level with which this can

Table 3.1
Summary of Proposed Objectives

Pollutant	Standard concentration	Standard measured as	Objective – to be achieved by 2005
Benzene	5 ppb	running annual mean	**5 ppb**
1,3-Butadiene	1 ppb	running annual mean	**1 ppb**
Carbon monoxide	10 ppm	running 8-hour mean	**10 ppm**
Lead	0.5 µg/m^3	annual mean	**0.5 µg/m^3**
Nitrogen dioxide	150 ppb	1 hour mean	**150 ppb, hourly mean★**
	21 ppb	annual mean	**21 ppb, annual mean★**
Ozone	50 ppb	running 8-hour mean	**50 ppb, measured as the 97th percentile★**
Fine particles (PM$_{10}$)	50 µg/m^3	running 24-hour mean	**50 µg/m^3 measured as the 99th percentile,★**
Sulphur dioxide	100 ppb	15 minute mean	**100 ppb measured as the 99.9th percentile★**

ppm = parts per million; ppb = parts per billion; µg/m^3 = micrograms per cubic metre
★ = these objectives are to be regarded as provisional, as described in paragraph 21 above

be carried out. There is therefore a vital role for the local authorities. As part of the process of local air quality management, it will fall to them, acting within central guidance, to assess how far any recorded exceedences are significant in exposure terms and require remedial action.

Percentiles 23. In formulating the objectives, it is sometimes appropriate that, for a standard with a short averaging time, the objective for the pollutant in question should be expressed in terms of percentile compliance. It is an approach that has been used in setting European air quality limit values, and is applicable here. The concept of percentile compliance is explained in Box 3.B.

Box 3.B
Percentile Compliance

If the objective is to be complied with at the 99.9th percentile, then 99.9% of measurements at each measuring point in the relevant period (usually one year) must be at or below the level specified. Taking the example of 15-minute values, therefore, there should be 365 x 24 x 4 = 35040 measurements, assuming perfect operation of the monitoring station throughout the year. All but the highest 0.1% of measurements – in this case the highest 35 15-minute values – must be at or below the value specified.

The reason for taking the percentile approach is that there will always be the possibility of occasions when it would not be appropriate to try to prevent exceedences of the objective level. Such occasions may be when 100% compliance would require disproportionately expensive abatement measures; or may arise for social and cultural reasons which it would be inappropriate to control or ban – national festivals, for example, such as Bonfire Night; or may be due to uncontrollable natural sources or adverse weather conditions. For reasons of practicability and proportionality, therefore, these targets do not require complete compliance with the given concentration value.

Relationship between UK air quality objectives and EC limit values

24. There are currently EC limit values for nitrogen dioxide, lead, sulphur dioxide and particles. The UK's obligation to comply with those limit values remains unchanged. In due course, the European Commission will come forward with proposals for daughter directives under the newly adopted framework Directive, setting revised or new air quality limit values. The first group of proposals will cover the pollutants mentioned above for which EC limit values already exist. The UK will, of course, be bound by the new limits which are set in the light of those proposals, and the deadlines for their achievement. Under the terms of the Directive, the UK will in all cases need to demonstrate compliance with the limit values when they come into force. Where there is a difference between the European limit value and the UK's air quality objective the Government will review the technical basis for the objective and the requirements for demonstrating compliance with the limit value, and will take a view whether the national objective needs adjustment.

25. The Government is mindful of the fact that EPAQS has recommended a further standard for benzene of 1ppb, to be adopted in the longer term. Whereas the balance of factors to be considered currently militates against the adoption of this level as an objective for this Strategy, the possibility of moving to this level will be re-examined in the review of the Strategy in two years time. Fuller consideration is given in Part II Chapter II.3.

26. The Government notes the recommendation of EPAQS that the air quality standard for PM_{10} should be reviewed within five years. The Government will consider this in the first review of the Strategy and, in particular, will examine the possibility of developing standards for size fractions or alternative measures of particles other than PM_{10}. EPAQS has also recommended that the air quality standards for 1,3-butadiene be reviewed after a period of, at most five years; and that a long term standard for nitrogen dioxide should be reconsidered within three years. The Government will also consider these issues in the first review of the strategy.

27. The review of the Strategy will also include an examination of the possibilities of extending the scope of the Strategy to other pollutants, notably polycyclic aromatic hydrocarbons (PAHs) on which the Expert Panel started work at the end of 1996. PAHs have been identified as priority pollutants under the UNECE/LRTAP Convention's proposed Protocol on Persistent Organic Pollutants. This protocol is scheduled for agreement in 1997 and may contain general or specific provisions for controls on the emissions of PAHs.

28. The Government has decided to concentrate on the protection of human health in this Strategy, and therefore is not, for the time being, proposing separate air quality standards or objectives defined in terms of ecosystem protection. The measures proposed to meet the targets set will of course help to reduce the impact of emissions on the environment as well as on health. The extent to which ecosystem protection can be further integrated into the National Air Quality Strategy will be addressed in its review.

Conclusion

29. Standards and objectives relating to air quality are the fulcrum of this Strategy. The fundamental aim of Governmental air quality policy is to render polluting emissions harmless, and a strategy to achieve this aim needs first to define the levels of harmlessness, and then to direct its policy towards the achievement of those levels by means of objectives as costs and benefits dictate. The standards are based primarily on advice from EPAQS, and represent levels at which no significant health effects would be expected. It is expected that, for these pollutants, the Government will be able, on the basis of existing and identifiable additional policies, to steer ambient levels to meet the standard as closely as costs and benefits can justify by the year 2005.

Chapter 4: Achieving the Objectives

Introduction

1. Chapter 3 of this Strategy sets out the air quality standards and objectives which the Government has decided to adopt. This chapter looks at those objectives in more detail and, in particular, sets out the main policies and measures by which they will be achieved. These policies are then elaborated and developed, as they apply to the separate industry, transport and local government (including domestic sources) sectors in later chapters.

2. As described earlier, the Government's method has been to focus on key pollutants for which objectives should be set. These pollutants are considered in detail in separate chapters in Part II of this Strategy. The current levels of pollution are quantified and forecasts are set out of the future levels and trends on the basis of recent studies of sources and emissions. The analyses in these chapters take into account existing and likely policies and compare the outcome to the 2005 objective levels, to identify whether any policy gap exists which would require additional measures either at national or local level.

3. This chapter pulls together conclusions from those sections to identify the broad priorities; to distinguish those pollutants where additional measures are likely to be necessary and what they should be; and to consider how measures taken to address one pollutant may complement or interact with those for another.

Pollutant control requirements

4. The Government has first considered whether there are any pollutants where it can be confident that measures already in place will be sufficient to secure the objectives of the Strategy. On the basis of the analysis in Part II, it is concluded that this is the case for benzene, 1,3-butadiene, carbon monoxide and, except in a very small number of cases, lead.

5. In each of these cases the analysis suggests that the standard should be achievable everywhere within the UK by 2005. In certain cases, it could be achieved earlier. The Government's policy will, therefore, be to adhere carefully to those measures which are likely to bring about that outcome, and monitor closely to ensure that the expected progress is maintained. Clearly, some new measures introduced to secure reductions in other pollutants may further reduce levels of these pollutants. The Government will welcome this as an added bonus. To the extent that through the operation of pollution control based on BATNEEC (Best Available Techniques Not Entailing Excessive Cost) and other policies, the targets can be achieved cost-effectively before 2005, this too will be encouraged.

6. The analysis then points to a number of priority pollutants where further measures may be necessary to secure the Strategy's objectives. These are: sulphur dioxide, particles (as PM_{10}), nitrogen dioxide and ozone. Among the key elements of the Strategy are therefore the targets and core policies related to these individual pollutants.

7. A programme embracing the achievement of the targets for these four pollutants will tackle the UK's major residual air quality problems, substantially disposing of the long term historical problem of sulphur dioxide and effectively eliminating significant episodes of both summertime and wintertime smog.

8. The following sections set out the approach adopted in developing the policies required to achieve the objectives in each of the central areas identified above, and how it relates to the overall strategic principles. Key principles are set out for identifying relevant policies, and the policies implied by these principles are reviewed in the light of the criteria for each of the pollutants.

Principles for the selection of further measures

9. The choice of objectives takes into account the available evidence on practicability and the economic consequences of attempting the level of abatement required. Objectives will be achieved by a number of abatement measures, depending on the scale of the reductions required and the nature of the pollutant. Such measures are not necessarily "end-of-pipe" technological fixes, focussed on single pollutants, and some will impact on several polluting substances; similarly for most substances more than one policy measure will contribute to abatement. For example, the first four substances in Table 3.1 will be abated significantly by measures taken under the Vehicles Directives, but the Stage I Volatile Organic Compound (VOCs) controls Directive, implementation of Integrated Pollution Control (IPC) at stationary sources, fuel taxation and other measures affecting overall vehicle usage will also contribute. Several of these pollutants are abated by measures which are directed principally at other kinds of damage - for example, in the case of SO_2, damage to buildings, crops, forestry, land and surface water ecosystems.

10. Abatement measures, whether taken under domestic legislation, or in accordance with European directives and other international agreements, are subject to careful assessment of the costs of compliance. The impact of proposed measures on the sectors involved is assessed together with consideration of benefits. The process involves extensive consultation with affected parties, and, where relevant, consideration of a wide range of abatement options.

Costs and benefits

11. Where supplementary measures are indicated, these are to be subject to an assessment of the costs and benefits before the need to adopt, given the time-frame of the Strategy, becomes evident. It is a fundamental principle of all Government policy that measures which incur a cost should achieve equivalent or greater benefits, and that the option taken could not be substituted by another which achieves the same benefit at less cost. This principle is of course subject to limitations, not least the fact that benefits are often hard to define in monetary terms, or that in some cases it would be inappropriate even to attempt to ascribe such values. There should, however, be an assessment of the merits of action and the options for action in terms of costs and benefits, as far as is possible and appropriate. In developing objective-led strategies, it is necessary to look across the various sectors for comparative assessments of the relative marginal costs and benefits of further action in each sector. This will be closely studied in the review of the Strategy, in the light of the uncertainties relating to these four pollutants, as identified in Chapter 3.

12. Some general comments are possible with regard to the different levels at which policies are introduced and therefore assessed. Where the principal measures for pollution abatement are developed on a Europe-wide basis, as in the case of vehicle emission standards, they are necessarily subject to an assessment of the costs and benefits on a

European scale. As part of this approach, the European Commission's proposals for new vehicle and fuel standards for 2000 follow an extensive cost-effectiveness study carried out by the European Commission together with the European oil and vehicle industries (the Auto-Oil programme). These are discussed in Annex 1. Where improvements are to be achieved through industrial upgrading, the cost of compliance is considered on a site by site basis. Some of the measures to alleviate pollution hotspots will necessitate action at a local level. This type of action will often affect considerations far wider than simply air pollution, and the overall economic effect will vary between different localities. In such circumstances, the balance of costs and benefits must be assessed at a local level.

13. With regard to the damage from air pollution, it is difficult to disentangle the effect of each individual pollutant. Hence, these costs are often best considered for air pollution generally. The Department of Health's latest assessment is that air pollution is at present responsible each year for several thousand advanced deaths; for ten to twenty thousand hospital admissions, and for many thousands of instances of illness, reduced activity, distress and discomfort. It is less difficult to estimate the monetary costs involved in related hospital visits and admissions, consultations and medication.

14. It is more difficult to assess people's willingness to pay to avoid the suffering, distress and inconvenience caused by any ill-health from air pollution, and this is likely to form a substantial part of the costs of air pollution. Ascribing values to non-treated illness, reduced activity or discomfort, and advanced death is contentious and, despite a growing focus on these issues in the literature, such values are still at best highly subjective. The same is true for assessing the distress caused by dirt and odour which have no significant physical effects. However, many pollutants also cause measurable and considerable damage to structures, crops, commercial forestry and fisheries, and sensitive eco-systems. Monetary valuation of this damage, and hence of the benefits of reducing pollution is less difficult, although still challenging. Annex 1 to this Strategy provides an analysis of the work which has been carried out on the estimation of damage cost estimates. It will be a priority of the preparation of the first review of the Strategy to develop understanding of the costs and benefits of air pollution control measures. In this context the Government is committed to undertaking the research and economic analysis that will be required to underpin the review of the Strategy and to set targets which are firm rather than provisional.

15. It should be recognised that there can be direct benefits to industry from the introduction of new or more stringent environmental standards. There is a growing industry in manufacturing and exporting environmentally-friendly goods. Moreover some requirements can reduce costs. For example, in the vehicle refinishing industry the required use of high volume low pressure (HVLP) spray guns, in order to reduce VOC emissions, has decreased the amount of paint which is lost in the spraying process and therefore represents a cost saving to the companies.

The levels of action   16. The Government believes that the management of the quality of air in the UK should operate at two levels. In the first instance, policies to reduce emissions and to engineer the lowering of ambient levels should be developed and applied, as far as is practical and cost-effective, universally. This encompasses both national air quality policies and the implementation of European legislation and other international commitments. The aim of these measures should be, in lowering ambient levels as far as possible within cost-benefit criteria, to provide the platform for any supplementary action which may be needed to tackle pollution hotspots. It is expected that universally applied policies should be sufficient to achieve the air quality targets contained in this Strategy for a significant proportion of the country, and in the case of some pollutants, for the country as a whole. The role of the system of local air quality management which was introduced by the Environment Act 1995 is, inter alia, to provide a fine tuning or corrective system where existing central policies are not sufficient to meet the targets, and further national measures would be too blunt or too expensive an option.

17. **Technical controls.** Although considerable progress has been made, some gaps remain to be filled and the Government must consider the full range of policy options available, including both process controls and economic instruments. Any effective strategy for improving air quality will need to be anchored at first on those instruments which deliver the most substantial reductions in emissions with a high degree of certainty in achievement and of enforceability. This, of course, is subject to considerations of cost-effectiveness. The importance of scale and certainty means that the Government must look in particular to what is achievable through technical controls on sources - in particular, the contribution from upgrading programmes for industrial sources and the setting of vehicle emission and fuel quality standards. This, however, does not preclude early consideration of complementary or even alternative measures.

18. **Economic instruments.** Many of the detailed control strategies outlined in the rest of this document are based upon regulation. The Government recognises, however, that economic instruments should in theory allow firms and individuals greater flexibility in achieving particular aims. In practice the issues are complex, and the choice of instruments must be examined on a case-by-case basis. Furthermore, there is no supposition that either an economic instrument or a regulation are best on their own - a mix of the two can often be more effective.

19. A number of economic instruments in the field of air quality policy already exist. The differential between the rate of duty on leaded and on unleaded petrol has helped secure a 70% reduction in airborne lead. Successive and pledged future increases in petrol and diesel duty have promoted carbon dioxide savings and reductions in polluting emissions. Most recently, the 1996 Budget introduced a further significant reduction in the duty on road fuel gases, a lower duty on ultra low-sulphur diesel relative to ordinary diesel, and announced the intent to introduce an incentive for low emission technologies on heavy goods vehicles.

Emission reduction strategies
Nitrogen dioxide (and nitrogen oxides)

Objective: to achieve the standards of 150 ppb hourly mean, and of 21 ppb, annual mean, by 2005

20. Total emissions of oxides of nitrogen peaked towards the beginning of the 1990s, and are now beginning to fall. The UK committed itself under the first UNECE NO_x Protocol to return its emissions to 1987 levels by 1994, and has achieved this target.

21. Further reduction of NO_x emissions would help meet two air quality targets - the lowering of ambient levels of nitrogen dioxide (NO_2), and the containment of ozone formation. The general policy objective for nitrogen dioxide is to reduce ambient levels to the extent that annual average levels are kept continuously low and peak episodes of wintertime smog are avoided. Ambient levels of NO_2 occasionally breach or threaten to breach the hourly air quality objective level, particularly in urban areas. In such areas it is road transport which is the major contributory cause, although other point sources can play a significant part. It is estimated that emission reductions in the region of 48-62% on 1995 levels would be needed to secure the consistent achievement of the objectives by the year 2005. The projected growth in road traffic and the vehicle fleet over the next twenty to thirty years will lead to a rise in the level of emissions after the gains made through the complete penetration of catalytic converters.

22. The principal means of achieving the necessary reduction must be the vehicle emission and fuel quality standards which are to be negotiated in the European Community. The standards already in place and other abatement measures are likely to achieve reductions of around 42% in urban road transport-derived NO_x emissions, relative to 1995 levels, by the year 2010. It appears that the Commission's new proposals, arising out of the Auto-Oil programme would, when added to the measures in place, achieve reductions of around 45-47% in urban NO_x emissions by the earlier date of 2005. This is the bulk of the reduction required, but may still not be sufficient to meet the objective everywhere. It is estimated that there will be a shortfall in the forecast emissions reduction of around 5-10% or more. The application of pollution control to industrial sources, by both the Environment Agencies and local authorities, will be essential in the areas where stationary sources are making a significant contribution to local ambient levels. Elsewhere, in urban areas, locally applied measures, particularly traffic management, along the lines described in Chapters 6 and 7, may well present the best opportunity for tackling local hotspots, although their impact is difficult to quantify at this stage.

23. Air pollution episodes are seldom characterised by one pollutant alone, and, in particular, particles and nitrogen dioxide are commonly associated in a typical winter air pollution episode in the UK. However, the limited data available suggests that the effects of nitrogen dioxide lead to fewer advanced deaths than PM_{10} - perhaps some hundreds in a typical year. This information will be supplemented by a European-wide study, the APHEA project, which is nearing completion. The study examined the effects of various air pollutants on health, including nitrogen dioxide, and will allow the impact to be quantified much more accurately than previously possible. The abatement costs for transport are very similar to those for PM_{10}. In particular, it is estimated that the package of vehicle measures mentioned above would lead to significant reductions in vehicle emissions. In the case of industrial upgrading, the operation of pollution control will ensure that the pace of implementation does not entail costs disproportionate to the health and environmental gains.

Ozone   *Objective: to achieve the standard of 50 ppb, measured as a running 8 hour mean, 97th percentile[1] by 2005*

24. Ozone is an aggressive pollutant which damages crops and materials, and can cause health effects, in particular for those with respiratory problems. A standard of 50 ppb, 8-hour running mean as recommended by EPAQS was intended to provide a margin of safety below the level of around 80-100 ppb at which discernible effects are observed. EPAQS recognised the challenging nature of this standard, and that at times in the winter natural fluctuations in the stratosphere could cause it to be breached. EPAQS also recognised that in the summer months a significant contribution to the observed concentration was from polluted air arriving from the rest of Europe.

25. The Government's aim is to move progressively towards the ozone standard, or as quickly as the balance of costs and benefits justifies by 2005, so as to secure by then, in effect, substantial elimination of significant summertime smog episodes. Confidence in the degree to which this aim and the 2005 objective can be met is influenced by a number of issues. First it is clear that, however severe a reduction in UK emissions of ozone precursors could be established during the summer months, these would not be sufficient to reach the standard unless there was a commensurate effort throughout Europe to reduce emissions. It is difficult to estimate the reduction in pollution expected elsewhere on the continent over the next ten years from national and international measures. Summertime ozone incidents in Northern Europe are at least as severe as in the UK, and a number of national programmes are in place or planned. For example, the eight countries of northwest Europe[2] which attended the Ministerial Conference on Tropospheric Ozone in Northwest Europe in May 1996 committed themselves to taking measures designed to eliminate ozone episodes[3] within the region by 2005. However, a number of the UNECE Protocols have yet to come into force, and agreement has yet to be reached on a number of key EC directives. These international instruments target ozone precursors in general and do not focus especially on those with the most significant photo-oxidant potential. The European Commission has just started work on preparing an EU-wide ozone strategy, to which the UK will be contributing in detail.

26. Over the next few years the Government's strategies for VOC and NO_x abatement will deliver significant reductions in emissions of these major ozone precursors. Through the application of the upgrading plans of Integrated Pollution Control (IPC) and Local Air Pollution Control (LAPC) and through the introduction of new vehicle and fuel standards, the Government expects that the reduction in total VOC emissions will be almost 40% in 2005 measured against 1988 levels. Similarly, emissions of NO_x are expected to have fallen by 40-45% over the same period. These are measures which apply throughout the year, and offer additional improvements in general air quality. Measures put in place since the 1980s appear to be having an effect, with evidence of a gradual reduction of ozone incidents over the decade.

27. Methods of estimating the full effect of these reductions on the scale of summertime ozone incidents are still in their infancy. However, despite

---

[1] 97th percentile as defined in Chapter II.8.

[2] Belgium, Denmark, France, Germany, Ireland, Luxembourg, the Netherlands and the UK.

[3] Based on the EC Information Threshold from Directive 92/72/EEC (the Ozone Directive).

the considerable scientific uncertainties that remain, it is clear that summertime incidents in the UK cannot be contained unless there is action across Europe. Furthermore, the necessary reductions are unlikely to be found solely from simple, across the board reductions in annual emissions. Further measures targeted at key pollutants and at critical times at the onset of an incident would seem to be necessary, though some of these could be achieved through well-timed short term voluntary measures. With these additional actions, it may be possible to achieve the standard by 2005 throughout the summer for much of the UK, and in the South for all but 10 days or better on average. These substantial reductions mean that it would be unlikely that the EC Information Threshold (90 ppb, hourly average) would be breached.

28. **Accordingly the Government believes it is right to set the ambitious policy objective of achieving the standard, on a 97th percentile compliance, in the UK by 2005.** However, the substantial uncertainties that remain, not least action in the rest of Europe, mean that this objective may need to be revised. On the basis of policies already in place it is expected that the standard will be breached on only a few days in the summer, and that ambient air quality will not be sufficient to trigger the EC Information Threshold, except in exceptional weather conditions. In seeking further reductions the Government will:

■ seek early agreement and ratification of a multi-pollutant UNECE Protocol;

■ seek to ensure that the proposed EC Solvents Directive gives appropriate priority to solvents with high Photochemical Ozone Creation Potential (POCP);

■ improve coordination with neighbouring states on warnings on potential ozone episodes; and

■ discuss with industry and other interest groups possible voluntary measures to limit emissions during the onset of photo-oxidant emissions.

29. It is difficult to quantify the monetary benefits of reducing exposure to ozone. A recent epidemiological study of ozone effects in the UK (part of the wider APHEA study of air pollutant effects on health) reported for the first time, associations between ozone levels and mortality. The extent of these potential effects, in terms of the extent of advancement of death or numbers of people affected is not clear at this early stage, but the Government believes that these results should be taken seriously.

30. There are also well established relationships between elevated ozone levels and impairment of lung function and increases in hospital admissions, published by the WHO in its revision of air quality guidelines. In addition to these effects on mortality and hospital admissions, it is important also to consider the other effects, for example on lung function, which individually may be small, but may affect large numbers of people and therefore have a large effect on public health. There are relatively few assessments of the monetary cost this represents, and the uncertainties involved mean that any figures must be treated with due caution.

31. The benefits of ozone reductions will also be felt through reductions in damage to materials and buildings. Accurate figures for the UK are difficult to obtain and work is under way to derive data, but a recently published estimate suggests that damage costs are up to £350 million

annually. Ozone reductions will also benefit crop yields and tree health. The extent of these benefits is currently under discussion within the UNECE and the EC.

32. It is, therefore, not a straightforward task to evaluate the benefits of ozone reductions. Quantifying the costs of the technological options for abatement is more straightforward. It is, however, less clear to what extent these costs can be set against benefits from ozone reductions, given that the majority of technical measures will reduce emissions both of pollutants which have adverse effects in their own right, as well as being ozone precursors, and of other harmful pollutants. Moreover, given the scale of the ozone problem and the sources involved in its formation, it is more appropriate to consider these issues on a European scale.

33. Strategies to reduce ozone concentrations will therefore be pursued on a European scale, taking due account, within the limitations discussed above, of the costs and benefits.

Particles (PM$_{10}$)   *Objective: to achieve the standard of 50 $\mu g/m^3$, measured as a running 24 hour mean, 99th percentile,[4] by 2005*

34. In November 1995, following publication of the EPAQS and Committee on the Medical Effects of Air Pollution (COMEAP)[5] reports on particles, the Government issued its preliminary policy proposals for securing the EPAQS recommended standard by 2005[6]. The Government has now considered the issues further in the light of the Quality of Urban Air Review Group (QUARG) report on particulate levels[7]. That report concludes that:

"Detailed analyses....clearly show the immense importance of road traffic emissions in influencing....PM$_{10}$ concentrations when these are elevated. In winter, a dramatic reduction of hourly exceedences of 50 $\mu g/m^3$ of PM$_{10}$ could be achieved solely by limiting road traffic exhaust emissions.

It is clear from the data and analyses....that stringent additional controls on particulate matter from road transport and upon the emissions across Europe of sulphur and nitrogen oxides responsible for the formation of secondary particles in the atmosphere, will be essential if UK urban particulate matter concentrations are to be reduced in line with the recommendation of the Expert Panel on Air Quality Standards"

35. The general objective is to reduce PM$_{10}$ levels in towns to such an extent as to constrain their contribution to potential wintertime smog episodes, and to secure a steady reduction in annual average levels, so as to minimise the risk to human health. Given the concentrations in urban areas and the contribution to those levels from road transport, the principal target of policy in this area must be the abatement of PM$_{10}$ emissions from urban transport sources.

[4] 99th percentile as defined in Chapter II.9

[5] an independent Committee of experts, reporting to the Department of Health.

[6] "Health Effects of Particles" The Government's preliminary response to the reports of the Committee on the Medical Effects of Air Pollutants and the Expert Panel on Air Quality Standards - DOE, DH and DOT, London November 1995.

[7] Quality of Urban Air Review Group: *Airborne Particulates in the United Kingdom,* 1996

36. To meet the objective for PM_{10} in 2005 throughout urban areas (where levels are highest), it is estimated that a reduction of around 60% on 1995 emissions will be required. The key measures to be adopted, set out in the Government's November 1995 statement, are:

- to press for a cost-effective package of vehicle and fuel standards in negotiation of the European Community's post-2000 standards, in order to secure a significant reduction in emissions;

- to provide appropriate guidance to local authorities on their duties under the Environment Act 1995, with regard to traffic management in pollution hotspots;

- to reinforce measures to ensure compliance with vehicle emission limits;

- to complete investigations into the potential of alternative fuels, and to encourage the development of those which are shown to provide the opportunity for cost-effective improvements in air quality;

- to keep under review the use of fiscal incentives as a means of encouraging environmentally friendly transport choices;

- to continue to apply the principles of pollution control to industries regulated by the Environment Agencies and local authorities under the Environmental Protection Act 1990 (EPA 90); and

- the measures introduced in the 1996 (and earlier) Budgets which will contribute to reductions in particle emissions.

37. Current indications are that the proposed measures will secure 50-55% reduction on 1995 emissions. This means that there will be a maximum shortfall of 5-10%, leaving some areas short of meeting the objectives. The extent to which the Budget measures will reduce this shortfall is not clear at present, but they will nonetheless make a contribution. Further measures will have to be implemented in those areas in order to make up this shortfall and meet the objective. The policy gap associated with PM_{10} is similar to that for NO_2 in that it is not yet clear whether the measures available will be capable of meeting the objective cost-effectively.

38. Current evidence suggests that PM_{10} is a major contributor to the effects of air pollution, in terms of hospitalisations and advanced deaths. In addition, PM_{10} makes a considerable contribution to the soiling of buildings. The benefits offered by achievement of that target will, therefore, be substantial. Some attempts have been made to quantify these benefits, citing significant sums, although considerable uncertainties remain as to their validity. Most of the abatement costs involved in achieving the target will arise from measures to improve the emissions performance of new vehicles, from 2000 onwards. The European Auto-Oil programme estimated the costs associated with achieving a package of reductions in emissions of various pollutants.

39. In view of the scale of the health effects, the Government will want to see the smaller contributory measures in terms of industrial abatement and traffic management introduced as soon as possible. In the case of industrial upgrading, the operation of BATNEEC will ensure that the pace of implementation does not entail costs disproportionate to the health and environmental gains. In the case of traffic management and other measures, it will be for local authorities to address in the normal way the

overall balance of costs and benefits. These are likely to include far more than simply costs and benefits from improved air quality.

Sulphur Dioxide

Objective: to achieve the standard of 100 ppb, measured as a 15 minute mean, 99.9th percentile[8] by 2005

40. Historically the UK's worst SO_2 problems were concentrated in urban smogs resulting from domestic coal burning and industrial combustion processes in urban areas. As is shown in Chapter II.9 this was substantially dealt with by changes in domestic fuel use and by shifting major industrial processes, in particular power generation, to large stations away from urban areas. That policy has yielded great benefits, but one consequence was that it did not resolve, and may even have exacerbated, long distance transport and deposition as acid rain. That is now being progressively tackled, and under the 1994 Oslo Protocol the UK is committed to reducing SO_2 emissions by 70% by 2005 and by 80% by 2010 against a 1980 baseline.

41. Achieving the objective for ambient levels of SO_2 may require a different focus than the attainment of national aggregate emission reduction targets. However, although exceedences are currently occurring throughout the UK as the result of plumes from large combustion plants, the measures planned to achieve the aggregate reduction targets will go far to secure the air quality objectives. Residual problems of two kinds remain which will require particular attention:

- background levels in local domestic coal burning areas during winter; and

- localised plume grounding from combustion sources which by 2005 may still be burning coal or oil.

42. The Government's general objective is, therefore, to secure reductions sufficient to tackle these residual background levels in domestic coal burning areas and to tackle localised plume grounding.

43. The ongoing change from coal to other fuels for domestic heating should progressively address the first problem with incremental effect. At present however there are difficulties in Belfast and in some areas of Northern England, particularly those where free coal continues to be available. In Belfast, the conversion of a major power station to gas will help substantially, but is unlikely alone to be sufficient. Higher sulphur coal is entering the market, and therefore regulations are being made to address locally the sulphur content of fuel and its sale. In relevant local areas the Government will monitor the continuing contribution from its smoke control policies through to the first review of this Strategy. If adequate reductions are not being secured, the Government will then consider the introduction of regulations or economic measures aimed at reducing the level of sulphur in domestic fuel.

44. Exceedences caused by plume grounding should rapidly reduce as large black fuel combustion sources, in particular power stations, are either closed or reduce their period of operation. Total sulphur emissions from the electricity supply industry (ESI) in England and Wales are now planned to reduce by 85% between now and the year 2005. Improvements should result from the Environment Agency's operation of BATNEEC

---

[8] 99.9th percentile as defined in Chapter II.10

elsewhere in the combustion plant sector. It is, however, more difficult to predict the rate of progress with these other sources. Accordingly, for any sources not sufficiently abated by 2005 to avoid exceedences caused by plume grounding, a minimum requirement is likely to be the operation of arrangements similar to those already proposed for some power stations. These involve the burning of low sulphur fuel at periods when climatic conditions seem likely to increase risks of plume grounding.

45. This may still leave a problem from small unregulated sources. The Department of the Environment is undertaking research on this issue and by the time of the first review of its Sulphur Strategy in 1998, will have considered how far the achievement of the UK's international obligations, and of the ambient air quality objectives for 2005, may require additional measures. It is likely, for example, that if smaller combustion plant do provide a potential contribution to exceedences of the objective, a revision of the guidance on chimney stack heights will be needed to ensure that emissions from new plant do not lead to high local concentration. Other control options are available, including controls upon fuel quality and economic instruments to influence fuel use.

46. Given that steps of this kind are available, in principle, to supplement the major reductions in emissions from the ESI, and with the prospect of further upgrading programmes for other relevant industries, the objective should, by 2005, be substantially achievable in a cost effective manner. The cost-effective application of BATNEEC to the upgrading programme will ensure that excessive costs in relation to benefits are avoided. Research indicates, however, that in monetary terms the primary benefits of achieving the objective will derive from the health care costs avoided when plume grounding is controlled further. Where this is required the use of lower sulphur coal should impose no significant cost penalty.

Chapter 5: Business and Industry

Introduction

1. As will be evident from the discussion of policy options for the main pollutants, business and industry have a key role to play in meeting the objectives of the National Air Quality Strategy. Industrial processes are the source of significant proportions of national emissions of some air pollutants, and, in some cases, of local exceedences of air quality objectives. Further reductions in industrial emissions will need to play a part, along with reductions from other sources, in meeting the aims of the Strategy. The majority of polluting industry is already covered by the established systems of Integrated Pollution Control (IPC) and Local Air Pollution Control (LAPC). The former is regulated by the Environment Agency (EA) in England and Wales, and by the Scottish Environment Protection Agency (SEPA). The latter is administered by local authorities in England and Wales, and by SEPA in Scotland. It is primarily through the continued operation of these controls that reductions in industrial emissions will be secured.

2. In addition to reducing its emissions, however, industry has other roles to play:

- industrial innovation will be important in securing reductions in pollution in the most cost-effective way, not just from industry itself but from other sectors responsible for pollution, such as transport. New products and techniques will emerge as markets respond to the challenge of new targets, not least in the environmental technology sector;

- industry's response to the challenge of improving air quality needs also to be seen in the context of the encouragement the Government is providing to firms to adopt demonstrably high standards of environmental management overall, through accreditation schemes such as BS7750 and the EC Eco-Management and Audit Scheme (EMAS); and

- industry can also contribute significantly to the reduction of air pollution by voluntary action, in areas such as transport and energy consumption. In these areas many companies are voluntarily leading the way. Some have converted their fleets to alternative fuels. Some provide free vehicle emission checks or promote car sharing. Others are reviewing the range of their operations to identify changes that can be beneficial in air quality terms, in particular looking at improved energy management and energy efficiency technologies such as Combined Heat and Power. The Government warmly welcomes such initiatives, and is confident industry will consider ways in which its voluntary contribution to securing cleaner air can be further enhanced.

Air pollution from industry

3. Industry is a significant emitter of air pollutants. Table 5.1 shows the proportions of national emissions attributable to industry. As ozone is a secondary pollutant, data for ozone precursors - nitrogen oxides (NO_x) and non-methane volatile organic compounds (NMVOCs) - is shown.

4. Clearly, industrial sources are the dominant source of some pollutants, such as sulphur dioxide; significant contributors of others, such as VOCs and lead; and less significant emitters of eg carbon monoxide, which is generated mainly from traffic. The relative importance of action by industry in meeting the Government's air quality objectives is, therefore, likely to vary according to the pollutant.

Table 5.1:
Industrial Emissions in
the United Kingdom

Pollutant	Total UK Emissions in 1995 (kilotonnes)	Industrial Emissions (kilotonnes)	Industry as % of total
Benzene	35	6.9	20
1,3-Butadiene	9.6	1.2	13
Carbon monoxide	5478	667	12
Lead	1492[1]	276[1]	18
NO_x	2293	852	37
Particles	232	135	59
Sulphur dioxide	2365	2112	89
NMVOCs	2257	1195	53

[1] tonnes

The Strategy for
Industry

5. Using powers under the Environmental Protection Act 1990 to control pollution from prescribed industrial processes, the Environment Agency, local authorities in England and Wales and SEPA in Scotland have an important role in the implementation of the Strategy. In granting authorisations these enforcement authorities already take into account the existing statutory air quality limits set by a number of EC Directives. They will now be required to take into account this Strategy. The objectives of the Strategy are of a different nature to the existing EC air quality standards. The EC limits apply now. In contrast, the Strategy's objectives are for achievement by 2005. The objectives will be met over this period, at most locations, by the application of normal plant upgrading under current integrated pollution control and local air pollution controls based on Best Available Techniques Not Entailing Excessive Cost (BATNEEC). There will be an opportunity to review this expectation and the objectives themselves in 1999.

6. However plant could be expected to improve on these emission standards if there remained a local air quality hotspot at the end of this period, assuming a proportionate benefit for the local environment. At some sites there may be multiple sources of the pollutant (eg. lead and nitrogen dioxide). Here, as previously, there is no expectation that industrial emissions should bear a disproportionate burden compared with other sources taking account of BATNEEC. The main purpose of the local air quality arrangements in the Environment Act 1995 is to ensure that local authorities can manage air quality in their areas and that sensible decisions can be taken in such cases.

7. Concern has been expressed that the inclusion of section 7(12)(g) in the Environmental Protection Act 1990, which adds the air quality objectives to the list of mandatory considerations for the authorising agencies, could be interpreted as going beyond the principles of the Strategy. The Government will ensure, either through the existing powers under the Environment Act 1995, such as direction to the agencies, or, if necessary, through primary legislation, that industry is not subjected to a disproportionate burden or one beyond that which would be expected through the normal operation of BATNEEC.

8. The Government's strategy in seeking to meet its air quality objectives is to seek appropriate reductions in emissions from each sector responsible for a pollutant in accordance with judgments of the extra marginal costs and benefits of securing such reductions from such sources. This means, for example, that further reductions will not, in principle, be sought in emissions of a particular pollutant from one sector if reductions of similar benefit to air quality can be achieved from another emitter of the same pollutant at lower cost.

9. This general approach will, of course, be complicated by practical factors. In the case of plant regulated under IPC, for example, the task of EA and SEPA will continue to be to find the Best Practicable Environmental Option (BPEO) for the environment as a whole, rather than simply the best option for air quality. And there are other constraints, such as the practicality of obtaining "least cost" reductions in some sectors, site specific factors, and the fact that pollution control measures often address more than one pollutant. But, faced with the challenge of ambitious air quality targets, the Government starts from the general principle that the reductions in emissions needed to enable its air quality objectives to be met (and, indeed, all its pollution reduction targets generally) should be sought from the most cost-effective quarter.

10. This philosophy has underpinned the Government's introduction of the regulation of industrial sources of pollution according to the concept of BATNEEC. Under BATNEEC, the test of whether costs are "excessive" or not should mean that abatement options whose costs are significantly higher than other means of obtaining equivalent environmental benefits are not pursued. In cases where European legislation sets binding limits, agreed derogations would of course apply.

11. This principle that reductions in polluting emissions should be sought from sources where they can be achieved most cost-effectively applies both to the choice as between reductions from different firms in the same industrial sector, and to the balance between different industrial sectors. The level of reduction in emissions to be sought from any particular firm, or any particular sector of industry, will therefore reflect, inter alia, the relative abundance of cost-effective opportunities for reductions in emissions by that firm or sector.

12. The Government also attaches importance to encouraging a variety of responses to the challenge of reducing emissions. End-of-pipe technology, process change and product substitution all have a role to play, and the aim of the Government's policy is to leave industrial operators with as much flexibility as possible so they can choose measures which reduce emissions at lowest cost to them. It is generally preferable that conditions in industrial pollution control permits are expressed as site specific emission limits, rather than as requirements to institute particular technologies. The Government believes that this approach will help to foster a healthy market in environmental technology and innovation, whereas rigid conditions requiring currently available technological solutions may stifle such development.

13. In line with the philosophy behind IPC and LAPC, such emission limits will continue to be set on a plant by plant basis, thereby enabling full account to be taken of the cost-effectiveness of the opportunities for abatement at the plant in question. It is possible, however, that there will

be cases where conditions prescribing technologies or technical measures are more appropriate than emission limits, particularly where, for practical reasons, an emission limit might prove difficult to enforce.

14. There is clearly a role, in principle, for economic instruments in enabling the Government to secure improvements in air quality whilst preserving the maximum possible flexibility for industry and others, who are best placed to do so, to decide how those improvements are to be met. Economic instruments such as pollutant taxes and permit trading are being explored in the context of the UK's likely commitments under an EC Directive on Solvents, for example.

15. This type of approach lends itself most readily to the pursuit of aggregate national emissions targets. It is less obvious what part, if any, economic instruments can play in tackling local exceedences of air quality standards. However, the Government has recently commissioned consultants to investigate whether such instruments might be used to improve ambient levels of sulphur dioxide in air.

Regulation under IPC and LAPC

16. The principal means by which air quality objectives will impact on industry is through industrial pollution control. The Environmental Protection Act 1990 (EPA 90) introduced a system of prior authorisation for polluting industry, replacing a patchwork of controls under the Alkali etc Works Regulation Act 1906, the Clean Air Acts of 1956 and 1968 and the Public Health Act 1936. For industry, the key changes were the introduction of permits containing explicit conditions supported by a rigorous enforcement regime, the public scrutiny involved in placing applications and authorisations on registers, and, for the largest plant, the unified controls associated with IPC. The continual operation, and further development of these regimes represents a key element of the Strategy, essential to the achievement of its objectives.

17. Under the EPA 90 regime, some 2000 of the most polluting processes with emissions to at least two environmental media are being made subject to IPC, whilst some 13,000 further processes whose polluting emissions are predominantly to air have been made subject to LAPC.

18. As mentioned above, these two systems of industrial pollution control are based on the concept of BATNEEC. Under BATNEEC, industry is required to take all steps (subject to a test of excessive cost) to prevent, minimise or render harmless releases of a series of prescribed substances listed in regulations, and to render harmless other releases. All of the key pollutants addressed by this Strategy are prescribed substances for release to air, with the exception of ozone, which is a secondary pollutant. The precursors of ozone, VOCs and NO_x, are both prescribed.

19. Under IPC, industrial processes are regulated according to individual, site specific judgments of what is BATNEEC and, in the case of processes likely to give rise to emissions to more than one environmental medium, what is the BPEO. The site specificity of BATNEEC judgments means that appropriate account can be taken of local factors, such as dispersion conditions at the site, peculiarities of site configuration or particularly sensitive local environmental receptors.

20. The requirement for a judgment of the BPEO in the case of most processes subject to IPC means that air quality may not be the sole criterion according to which decisions about permit conditions are taken. Indeed, the primary purpose of IPC is to facilitate consideration of the effect of industrial installations on the environment as a whole. However, air quality is clearly a major consideration informing the regulator's judgment, and where already existing statutory air quality limits are likely to be exceeded the regulator is required by the legislation to consider what contribution regulated industry might make to achieving them. This might involve, for example, where no derogations existed, tighter conditions on an individual plant than would have been imposed had BATNEEC and BPEO been the sole considerations.

21. Under the Environment Act 1995, EA and SEPA will, in the future, have to have regard to this Strategy in the performance of their pollution control functions. The Agencies will need to be involved particularly in the development of action plans in designated areas of local air quality management, where IPC processes contribute to the exceedence or risk of exceedence of an air quality objective.

22. Under LAPC the same BATNEEC judgments are made and site specific decisions are taken. However, enforcing authority decisions tend to rely heavily on the statutory guidance issued by the Secretary of State for generic industrial sectors, particularly in view of the more standardised nature of LAPC processes. While the aggregate contribution from industry cannot be estimated with precision, for some pollutants and some sectors it is likely to be substantial.

The EC Integrated Pollution Prevention and Control Directive

23. The principles which underpin IPC and LAPC are reflected in the EC's Integrated Pollution Prevention and Control Directive, which was adopted in late 1996. Like IPC and LAPC, the Directive requires site specific permits which take account of the characteristics of each plant, its location and the state of the local environment. The concept of Best Available Techniques (BAT) which the Directive requires permits to be based upon is very close to BATNEEC, particularly as techniques will not be considered to be "available" unless the balance between their costs and benefits is acceptable. As with IPC and LAPC, decisions as to what is BAT will be taken on a plant by plant basis by the competent authorities in each Member State.

24. The Government has until 30 October 1999 to adopt the necessary transposing measures. The legislation under which industry is regulated will therefore change, although the principles and concepts on which the system works will remain familiar. It is possible, however, that the advent of IPPC will affect the distribution of plant between LAPC and IPC. The first in a series of consultation papers will be issued by the Government shortly.

25. In keeping with its normal approach to the implementation of European legislation, the Government will implement the Directive so as to impose the minimum burden possible on those it regulates, consistent with retaining an effective system of industrial pollution control.

Industrial Pollution Control and the Government's Waste Strategy

26. Industrial pollution control and the Government's waste strategy are complementary. Measures which minimise waste often prevent and minimise pollution, too. Whilst the Government's overall strategy for waste will lead to less waste being generated, measures such as the landfill tax could, by shifting the balance of waste management between landfill and other options, such as incineration, have implications for the achievement of the Government's air quality targets. Continued regulation of incinerators according to BATNEEC standards will help to ensure that any increases which may take place in the quantities of wastes incinerated do not threaten the achievement of those targets.

Achieving the Strategy's Targets: **The Contribution of Industrial Upgrading**

27. The continuing operation of IPC and LAPC will, over time, lead to reductions in the amounts of key pollutants emitted to air by the industries those systems regulate. This will be a result of upgrading and other measures to limit emissions from existing plant in accordance with BATNEEC, and the replacement of old plant with new. Over time, further reductions in the emissions from both existing and new plant is likely owing to an expected reduction in the levels of emissions which can be considered BATNEEC as technological change makes improvements in environmental performance cost-effective.

28. Comprehensive information about the progress of IPC and LAPC in reducing industrial emissions is still being collected. The progress of IPC in England and Wales will, in due course, be visible from the Chemical Release Inventory (CRI) operated by the EA. However, because of the time lag between the issuing of an authorisation and the availability of monitoring data to the enforcing authority, and because in some cases complete monitoring and reporting arrangements have not yet been required, the information which has been published in the first two annual reports of the CRI is incomplete. But in due course, the CRI will provide a picture of the trend in emissions of pollutants from the industries

Figure 5.1: Projected sulphur dioxide emissions from ESI 1990-2010

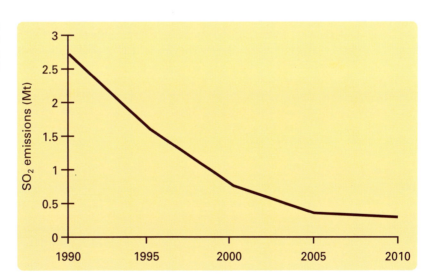

regulated under IPC. The EA will be developing the CRI so that it contains a full picture of emissions of the key pollutants from industry regulated under IPC.

29. It is possible, however, to see the beginning of a trend whereby IPC will result in significant reductions in industrial emissions. The first sector

of industry to be brought within IPC was the large combustion plant belonging to the electricity generators National Power and Powergen. Her Majesty's Inspectorate of Pollution (HMIP - now part of EA) recently varied the initial authorisations in order to tighten substantially the limit on the amount of sulphur dioxide the generators are allowed to emit. Figure 5.1 below shows projected emissions of sulphur dioxide from the Electricity Supply Industry (ESI) over the period 1990 - 2010. Since power generation accounts for 66% of national emissions of sulphur dioxide, a tightening of limits in this sector will have a marked impact on overall levels of sulphur dioxide in the ambient air.

30. All IPC and LAPC authorisations are subject to four-yearly review. Therefore, as the EA and SEPA review and, as appropriate, tighten IPC authorisations, and as local authorities in England and Wales and SEPA in Scotland review LAPC authorisations to reflect BATNEEC, further reductions in permitted industrial emissions of all key pollutants addressed in this Strategy can be anticipated in due course.

31. LAPC is also bringing about improvements in industrial emissions which will make a significant contribution to achieving the Strategy's objectives. In accordance with timetables for upgrading laid down in Secretary of State's guidance notes, the majority of initial improvements will take place during the period 1996 - 1998. These improvements will reduce emissions of a wide range of air pollutants.

32. For example, the emission limits specified in the Secretary of State's guidance note for cold blast cupolas should reduce emissions of particles from this type of plant by 80%. LAPC is also one of the tools by which the Government expects to have achieved the target in the UNECE VOC protocol of a 30% reduction in emissions of VOCs between 1988 and 1999, primarily through the controls in place on solvent-using industries, such as printing processes and coating processes. Industrial emissions of VOCs are expected to have fallen by around 35% between these dates.

33. While it is difficult to estimate its overall effects with precision, the progressive extension of industrial upgrading programmes should therefore represent, with new European vehicle and fuel standards, one of the two most substantial contributions to the Strategy's targets.

Sectors of industry not regulated by IPC or LAPC

34. In the case of some key pollutants there is a sizeable gap between the contribution of industry as a whole, and that of those installations already regulated under IPC and LAPC. As part of its strategy, the Government will consider whether reductions in key pollutants should be sought from industry other than that already regulated. An example might be combustion plant which is currently not regulated under LAPC (ie less than 20 MW rated thermal input). Contributions from smaller industrial enterprises of this type might be achieved by bringing them within the scope of industrial pollution control, or perhaps through alternative means, possibly involving an economic instrument such as a tax or permit trading scheme. Any extension of the scope of pollution control regulation would, of course, need to be subject to an assessment of the costs and benefits.

Industry's own interest in higher environmental standards

35. Measures to protect the environment will often involve substantial investment on the part of industry. The benefits are largely external, in the form of cleaner air, water and/or land. But industry can benefit financially from the process, too. For example, a plant regulated by HMIP (as it then was) under IPC was found to be burning oil incorrectly, resulting in the release of unburnt oil to air. The company saved money as a result of HMIP requiring it to ensure full combustion, which resulted in a lower consumption of oil. Good energy management and the implementation of energy efficient measures can deliver substantial savings on energy bills. Energy efficient technologies such as Combined Heat & Power (CHP) require a capital outlay but pay for themselves, usually within a relatively short period. Good solvent management is another example of how reductions in pollution can mean reductions in raw materials costs. Another example is the vehicle refinishing sector, regulated under LAPC, which has reaped significant savings from switching to more efficient paint spraying equipment which also reduces emissions of VOCs.

36. Industry also benefits in less tangible ways from its own better environmental performance. As trade becomes more global, businesses are finding it increasingly difficult to compete solely on price with suppliers from abroad. Customers, however, are becoming more sophisticated and are increasingly inclined to take into account the effect of a manufacturer on the environment when deciding which product to buy. And financiers and insurers, concerned about potential environmental liabilities, may offer better terms to companies which can demonstrate that they have these risks under control.

37. It is for this reason that the Government is encouraging industry to act in its own best interest, by improving and demonstrating its environmental credentials. The UK developed the world's first national standard for environmental management systems - BS 7750 - and has played a key role in helping the EC to design the Eco-Management and Audit Scheme (EMAS) which became fully operational in April 1995. The key features of EMAS are a properly audited environmental management system (such as BS 7750), and an independently verified public environmental statement.

Industry's Role in the Strategy

38. Over the period to 2005, the Government's aims for industrial air pollution control will be:

■ to put in place, for industry regulated under IPC or LAPC, a comprehensive set of improvement programmes leading to protection of the environment through the use of BATNEEC, drawing on the experience gained under IPC in the development of the ESI programmes and under LAPC those now established for a variety of other industrial sectors;

■ to improve the availability of information about the cost of securing reductions in pollution from different sectors, so as to enable decisions to be taken with sufficient confidence as to which sectors can and should be required to reduce emissions further;

■ to ensure that the Strategy is implemented so as to minimise the burdens on business and industry, consistent with meeting the air quality objectives;

■ to develop the CRI for IPC so that by, or soon after, the first review of this Strategy it can quantify how far the substantial expected further contributions from major industry to the Strategy's objectives are being secured;

■ to continue to explore the potential role for market instruments, such as taxes or permit trading, in the control of industrial and other emissions;

■ to implement the EC Directive on IPPC so as to build on the sound system of industrial pollution control already in place in the shape of IPC and LAPC;

■ to develop appropriate mechanisms to ensure that industry can play its part in achieving the benefits to be obtained from local air quality management; and

■ to look to industry, by its own voluntary measures, increasingly to become a partner with central and local government in securing the benefits which achievement of the objectives of this Strategy can offer, both to industry and to the community at large.

Chapter 6: Transport

Introduction

1. Air pollution in the UK has traditionally been associated with industrial activity and with the domestic burning of coal. These remain important. However, the application of best available techniques not entailing excessive cost (BATNEEC) as described in the previous chapter has led to sustainable reductions in emissions from these sources and shall continue to do so in the future. The position with transport is more complicated. In recent decades, transport emissions have grown to match or exceed other sources of many of the most important pollutants. In many areas, particularly urban, they have become the dominant source of air pollution emissions. Transport has also become one of the largest sources of greenhouse gas emissions and there is considerable overlap between air quality and climate change policy in this area. The growth in transport pollution is not, of course, due to less stringent standards for vehicles, but to the increase in traffic on our roads.

2. At the same time transport is an essential part of a thriving economy. A balance therefore needs to be found. This must ensure that the country has the modern transport system it needs to achieve sustainable economic growth, but with as little adverse impact as possible on the environment. This must be achieved in a way that ensures personal safety and allows freedom of choice, and at a cost it can afford. Considerations of environmental protection, personal safety and freedom of choice all point to the need to build effective protection of air quality into transport policies.

3. Policies for progressively abating such emissions must be at the heart of any effective national air quality strategy, and must embrace both measures to reduce emissions from individual vehicles and to reduce the quantity or growth in road transport in particularly sensitive areas. These are the central objectives of this part of the Strategy.

4. The role of future transport policy in addressing the problem of air pollution was also covered in the Government's Green Papers *Transport: The Way Forward*[1] and *Keeping Scotland Moving*[2]. Both the Green Papers described the weight which the Government attaches to the reduction of vehicle emissions, within the wider context of reducing the overall environmental impact of transport. The Government also published at the same time 'A Transport Strategy for London' setting out policies for the capital.

5. This Chapter first sets out projected trends in transport emissions and the reductions required to ensure that transport makes its contribution to achieving the targets and objectives of the Strategy. It then sets out the Government's key strategic aims in relation to transport and air quality, and the principles underlying them. The Chapter then moves on to describe the measures currently being taken, and those which it will, if necessary, consider, to ensure that the Strategy's transport objectives can be met. Most of the discussion is on road transport, but concluding sections look at air, rail and shipping.

Trends and Objectives in Road Transport

6. Between 1984 and 1994 the number of cars increased from just over 16 million to about 20.5 million, and the distance they drove increased significantly and is projected to continue growing. Figure 6.1 shows overall traffic growth estimates for the years 1965 to 2025.

[1] *Transport: The Way Forward* a Green Paper for England and Wales, was published in April 1996.

[2] *Keeping Scotland Moving* a Green Paper for Scotland, was published in February 1997.

Figure 6.1:
Motor vehicle traffic
1965-2025

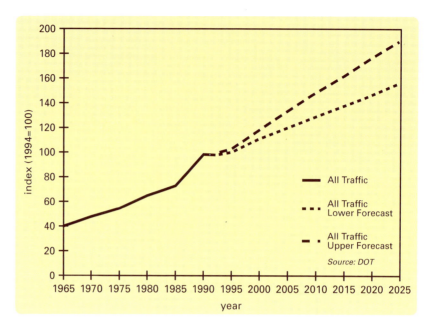

7. Until the 1980s, the increase in road transport was accompanied by similar increases in the main air pollutants. However, the introduction and promotion of unleaded fuel, and tighter vehicle standards, in particular the introduction of catalysts, are reversing this trend. As cleaner vehicles enter the market place over the next decade, these improvements should outweigh the effect of continued growth.

8. At present, however, as the analysis in Chapter 4 indicates, emissions from road transport, account roughly for the emissions shown in Table 6.1.

9. As the table indicates, the proportion of emissions which originate from transport varies from area to area, depending on the other local sources, the amount and type of traffic and the degree of congestion. Moreover, the source of pollution can vary considerably from these figures during a smog episode. For example, during winter episodes in most major UK cities the dominant contribution to PM_{10} levels comes from road traffic (notably from diesel powered vehicles), increasing from about a quarter to in excess of 70%.

Table 6.1
Contribution from Road
Transport to UK
Emissions

		1995 National emissions (k tonnes)	Contribution from road transport	
			% of national emissions	% of emissions in London
Benzene★		39	67%	not available
1,3-Butadiene★		10	77%	not available
CO		5478	75%	99%
Lead		1.47	78%	not available
NO$_x$		2295	46%	76%
Particles	PM$_{10}$	232	26%	not available
	Black smoke 356		50%	94%
SO$_2$		2365	2%	22%
VOC		2337	29%	97%

★ **1994 estimates used**

10. Figure 6.2 shows projection of vehicle emissions of NOx and particulates through to 2025, assuming existing policies only. The projection is based upon the mid-point of the low and high forecast of traffic growth. Assumptions are built into the model concerning the make-up and turn-over of the vehicle fleet, and the average trip length. It is assumed that driving conditions will remain constant. The relevant emission factor for each vehicle type is calculated from measurements of emissions from on-road vehicles under different driving cycles and conditions.

11. On present policies and technologies the expectation is that emissions will continue to decline to about 2010 but begin to rise again thereafter as the growth in road traffic offsets the reduction in emissions from individual vehicles.

12. In these circumstances, government policy must have two aims:

■ to ensure that the reductions in emissions predicted are realised, and where practicable accelerated, in order to help secure the air quality objectives for 2005; and

■ to sustain levels of emissions at the levels recommended as national air quality objectives once they have been reached.

**Figure 6.2
Future UK Urban Road
Transport Emissions of
Nitrogen Oxides**

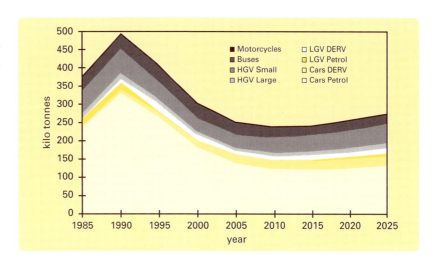

**Figure 6.2
Future UK Urban Road
Transport Emissions of
PM$_{10}$**

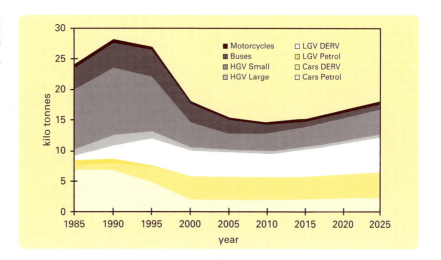

13. Given that vehicles are a major source of some priority pollutants, the achievement of the air quality objectives will require substantial reductions from the transport sector. However, the Government does not propose to set sectoral targets for the reduction of emissions. They could inhibit flexibility and conflict with the aim that reductions in emissions should be achieved by the most cost effective route across the sectors. In addition, sectoral targets ignore the variation in emissions profiles of different regions and locations. Therefore, there will be no separate target for the transport sector. The final package of measures must depend on detailed assessment of the cost effectiveness of individual measures within it and on the content of local air quality management plans. However, the Government estimates that meeting the air quality objectives in urban areas would imply a reduction in emissions of around 45%-60% for nitrogen oxides (NO_x), 60% for PM_{10} and a similar figure for volatile organic compounds, in relation to regional ozone concentrations.

14. The measures outlined in this chapter suggest ways in which reductions of this order for these pollutants may be achieved by the target date of 2005. Although the objectives are likely to be met in much of the country through national measures, there will be areas where this is not the case. Until recently, local and voluntary measures to reduce vehicle emissions have not been widespread. It is therefore difficult to quantify the emission reductions which may be achievable through such measures. As local traffic management measures are refined in the light of local authority experience, further evidence about the effectiveness of such schemes will be disseminated and fed in to the review of the Strategy in 1999.

15. In line with the principle of "polluter pays", it is the Government's view that transport users should bear the costs that are imposed by vehicle use upon the wider society. Accordingly, when opportunities arise and it is practical to do so, the Government will, in principle, look favourably on measures which help to internalise the costs of vehicle pollution and other environmental effects .

A Strategy for Transport and Air Quality

16. In *Air Quality: Meeting the Challenge* the Government set out the key principles that it would follow to secure reductions in air pollution:

■ improvements in vehicle and fuel technology to reduce emissions;

■ tighter controls on the existing vehicle fleet, its management and operation;

■ development of environmental responsibilities by fleet operators, particularly public service fleet operators, and by the public at large, in transport and vehicle use; and

■ changes in planning and transport policies which would reduce the need to travel and reliance on the car.

17. It is the Government's view that an effective strategic policy must incorporate all these four elements. At the heart of the policy is the need to balance them in the most cost effective manner. It must be recognised however that they operate over different timescales. It requires time for changes in vehicle technology to be applied, normally at least five years, although changes may in some cases be achievable over a shorter timescale. Adjusting transport systems or traffic management priorities in large and dense urban areas could typically take 5-10 years. Changes within the planning systems can take even longer to show significant results, and

it may only be some years into the next century that current changes in policy will have effect. For short term improvement, it is necessary to look at vehicle inspection and maintenance and at driver behaviour.

18. Given these differences in timescale, practicality and cost effectiveness, the Government proposes that the central elements of its strategy for transport emissions should operate as follows:

- the main contribution, to securing the necessary reduction in polluting emissions, will come from improvements in vehicle technology and fuels. The Government looks to achieve significant reductions in NO_x, PM_{10} and VOCs as a result of the Auto Oil 2000 European standards, and in line with the general principle of cost effectiveness;

- action will also be taken to encourage the use of less polluting alternative fuels, particularly in urban areas;

- for the residual reductions necessary beyond those which technology and fuel changes will yield, the Government will look to tighter enforcement of standards, and to local action on traffic management and improved driving practice. The Government will seek, wherever possible, to bring such measures into play quickly, to secure earlier improvement than otherwise might be achieved by relying on improved technology alone; and

- in the longer term, planning and other measures will help to reduce the need to travel and reliance on the car. This should help to avoid the prospects of an upturn in emissions that might occur from about 2010.

19. Measures which help reduce the need to travel will also make an important contribution to controlling emissions of greenhouse gases from transport. However, it is possible that some measures to reduce emissions of local air pollutants could lead to higher emissions of greenhouse gases, especially carbon dioxide, by reducing fuel efficiency. Similarly, whilst increased diesel penetration would help improve average fuel efficiency and reduce carbon dioxide emissions, it would have an adverse effect on air quality. Policies must, therefore, strike a careful balance between global and local priorities.

20. The remaining sections of this chapter set out the policies the Government has adopted, or proposes to adopt, in each of these areas; and identifies those areas where further measures could be taken forward if future reviews of the Strategy show it to be necessary. However, the introduction of catalytic converters may have limited the scope of car manufacturers to reduce emissions of carbon dioxide and have led to increased emissions of nitrous oxide, another greenhouse gas. This is an area where there is an unavoidable trade-off between the attainment of local air policy objectives and global climate change objectives.

Vehicle and Fuel Standards

21. Emissions standards for vehicles and fuel standards will have a key role to play in further reducing total pollutants from petrol and diesel-engined vehicles. There is a good basis on which to build. Since the 1970s, progressively more stringent requirements governing emissions from cars, trucks and buses have been introduced into UK law. Catalytic converters can reduce emissions from petrol-engined vehicles by 80%-90%. They were introduced in 1993 and are now fitted to around 25% of cars.

22. Emissions of carbon monoxide (CO), total hydrocarbons and nitrogen oxide ($HC+NO_x$) and particles are limited for new vehicles by:

- Directive 70/220/EEC as last amended by Directive 96/69/EC for passenger cars and light goods vehicles; and

- Directive 88/77/EEC as last amended by Directive 96/1/EC for heavy duty commercial vehicles.

23. Cleaner road fuel has also contributed to reductions in vehicle emissions. Unleaded petrol was introduced in the 1980s and the maximum permissible levels of lead in petrol were reduced. Unleaded petrol now accounts for around 70% of sales, and this figure will continue to rise as newer cars enter the market. In October 1994 the maximum permitted sulphur content in diesel fuel was reduced by a third, followed by a 75% reduction in 1996 under Directive 93/12/EC. The Government will consider whether further reductions in the sulphur content of diesel and/or petrol would be cost-effective, as part of the measures to be agreed for the 2000 vehicle standards. Cleaner road fuel could also help to control greenhouse emissions by allowing the development of technologies which at the same time meet local air quality and global climate change objectives.

24. If no additional improvements were undertaken from road transport, levels of air pollutants would be expected to continue to decline on both a national and urban scale into the next decade. Eventually, traffic growth would overtake the benefits which arise from the measures already in place. Further measures which are now being considered which will take effect from the year 2000 onwards will extend this timescale.

Next Steps
25. Further progress will, however, be needed to meet the air quality targets for nitrogen oxide and particles in this Strategy. The position is similar for many of the other Member States of the European Union, and to address this problem the European Commission has issued proposals for revised vehicle and fuel standards to take effect from 2000, and indicative standards for vehicles from 2005. The indicative standards for 2005 will be subject to a review concluding in 1998, and will also encompass a range of measures to reduce emissions, undertaken on the same cost-effectiveness basis as adopted when developing the 2000 proposals. The proposed package has been planned to make a significant contribution to tackling levels of nitrogen oxides, ozone, and particles in Europe. The package also includes consideration of alternative fuels and technologies, such as on-board diagnostics. The 2000 proposals, if adopted, would, combined with existing standards which are still to have their full effect, achieve emission reductions in the region of 50% on 1995 levels by the year 2005.

- The Government is playing an active part in the negotiation of these proposals, and is looking for substantial reductions which will form the backbone to achieving the reductions in transport emissions needed to meet our air quality targets.

Choice of Fuel
26. The debate over the relative environmental impacts of petrol and diesel is a complex one. Compared to catalyst-equipped petrol cars, diesel cars emit less carbon dioxide, carbon monoxide and hydrocarbons but more particulates and nitrogen oxides. The Government will continue to publish the latest evidence appropriate to enable the public to make informed decisions on their choice of vehicle. Advice from the Department of Health's Committee on the Medical Effects of Air Pollutants is that there is no basis on health grounds for preferring diesel vehicles to petrol-engined ones. Indeed, the present balance of evidence would

suggest that, while there are environmental pros and cons of each technology, three-way catalyst equipped petrol vehicles are likely to be less harmful in terms of health effects than diesel vehicles in the urban driving environment. It was partly for this reason that the Government decided to change the duty treatment of diesel in 1994, which was hitherto more generous in terms of pence per litre than for unleaded petrol.

27. Through the life-time of this Strategy and probably well beyond, it must be expected that petrol and diesel fuelled vehicles will be the predominant mode of road transport. Nevertheless, the Government considers that alternative fuelled vehicles could make a significant contribution to reducing the air and general environmental effects of transport.

28. The Energy Technology Support Unit (ETSU) recently published a study comparing emissions from alternative fuels over the whole life-cycle of the fuel, considering production, distribution and use (see Box 6.A). This will be supplemented within the next 2 years by the results of two series of trials. One of these is comparing emissions and costs of using alternative motor fuels in a range of different vehicles with conventional fuels. The other focuses especially on establishing the size profile of particle emissions from different vehicles.

BOX 6.A *The ETSU Report compares emissions from Light Goods Vehicles given current technologies. Their findings indicate that:* **Compressed Natural Gas** *(CNG) emits extremely low quantities of the key urban pollutants, nitrogen oxide and particulates. The potential climate change impact of using CNG, including the effects of both methane and and carbon dioxide, is between that of petrol and diesel.* **Liquid Petroleum Gas** *also offers very low emissions of particles and low emissions of nitrogen oxides. Again, the climate change impact was between that of petrol and diesel.* **Electric vehicles** *produce zero (or nearly zero) emissions at the point of use, and hence can enhance urban air quality. In considering the overall emissions, the source of electricity is a critical factor. They can offer overall savings in emissions of particulates, but the position on nitrogen oxides is less clear. Emissions from* **bio-diesel** *depend critically on the methods of farming employed. Its use can lead to savings in carbon dioxide emissions of 60%, but emissions of particulates and nitrogen oxides are increased.*

■ It is the Government's view that Compressed Natural Gas and Liquid Petroleum Gas have considerable environmental advantages for use in urban fleets of Light Duty and Heavy Goods Vehicles, and in bus fleets. The promotion of these fuels, via fiscal incentives, has been a feature of the 1995 and 1996 Budgets. In 1995 there was a 15% cut in duty on road fuel gases and a further 25% in the 1996 Budget. As a result of these measures the price at the pump should now be cheaper than petrol and diesel, helping to offset the costs of conversion. Where conversion to gas is not a realistic option, the use of low emissions technology and improved diesel formulation should be considered. Reformulated fuels do not require vehicle conversions.

■ Further, 1996 Budget measures discussed below, are also aimed at influencing choice of fuel.

Chapter 6: Transport

Fiscal Incentives

29. Road fuels are subject to fuel duty and to VAT at 17.5% and the levy of both taxes has proven to be an effective tool for encouraging a more sustainable approach to motoring: it allows each road transport user to respond in a flexible way; and can be effective at promoting the wide uptake of fuels and technologies which have less impact on the environment. For example, the different duty payable on leaded and unleaded petrol has contributed to a rapid uptake of unleaded petrol, and levels of air-borne lead have dropped by up to 70% in urban areas.

30. In general, the Government considers that motorists should as far as possible cover the economic costs of road use, including the environmental cost. The 1996 Budget reinforced this message by increasing the tax (duty plus VAT) on petrol and diesel 3 pence per litre in line with the Government's commitment to raise road fuel duties by an average of at least 5% a year in real terms. The primary aim of these measures is to help us to meet our commitments to limit carbon dioxide emissions, but it should also contribute to reducing emissions of VOCs, NO_X and carbon monoxide. However not all fuels are equally polluting. The tax (duty plus VAT) on super-unleaded petrol was increased by 4 pence a litre from May 1996 in recognition that its use, in cars without catalysts, can lead to greater emissions of aromatic compounds such as benzene than from other petrols. The 1996 Budget also contained an important air quality package aimed at the encouragement of environmentally friendly options:

- in line with the Government's fuel duty strategy, a 3 pence per litre increase in fuel tax (duty plus VAT) on petrol and diesel;

- duty on ultra-low sulphur diesel 1 pence less than on conventional diesel. This measure will come into effect as soon as possible after May 1997, when the UK obtains the necessary derogation from the European Commision;

- duty on road fuel gases was cut by a further 25% following the 15% cut in the 1995 Budget, reflecting their significant environmental advantages for use in urban fleets of light duty and heavy goods vehicles; and

- the intention to provide an incentive of up to £500 for lorries producing very low particulate emissions, to be introduced in 1998. This will provide a valuable incentive for lorries which meet a very low emission standard. This might be met by the fitting of particulate traps, or for smaller lorries, by conversion to gas. The Government is currently consulting on the details of the practical application of this proposal, with a view to implementation in early 1998.

Emissions Tests and Vehicle Maintenance

31. Even the emissions performance of vehicles newly entering into service can quickly deteriorate as a result of poor maintenance or poor driving. It is for these reasons that the emissions performance of all except the oldest vehicles in the fleet is assessed by metered checks at annual roadworthiness or MOT tests. UK in-use standards were tightened by 20% for diesel engined vehicles and by nearly 25% for petrol engined vehicles from September 1995. Even more stringent in-use requirements were introduced for catalyst equipped vehicles during 1996. These make use of manufacturers' specifications as to the emissions capabilities of each individual model of vehicle in order to optimise in-use control and maximise air quality benefits

32. Regular annual tests to check the emissions performance of all, other than new vehicles, are essential, and provide an opportunity to inform the public about the actual emissions performance of their particular vehicle, and the need to stay in tune. Moreover, the data can then also be used more generally, as an accurate indicator of the condition of the vehicle fleet in general.

■ New generation emissions meters are currently being introduced into MOT centres, which will provide the customer with an automatic print-out of the vehicle emissions recorded. By 1999, these meters will be available at all MOT centres.

■ A research project will start shortly, collating information from a sample of the new emissions meters, distributed around the country. This information will then be used to form a better picture of the general condition of the UK vehicle fleet. The Government is also currently considering whether more widespread collation of the data would be advisable.

33. However, the Government recognised some time ago that these compliance tests need to be complemented by other reminders during the year to encourage motorists and vehicle operators to pay attention to proper engine maintenance. This is why the Department of Transport's Vehicle Inspectorate introduced a programme of checking compliance with emissions standards checks at the roadside. During 1995 the scale and visibility of these checks was increased, with a programme of high-profile emission blitzes. The Government now proposes that a more permanent framework of roadside emission checks should be put in place, which both allows those areas where emissions may be most harmful to be particularly targeted, and gives local authorities a role, as part of their wider remit on local Air Quality Management.

■ To complement the annual in-use tests, the Vehicle Inspectorate Executive Agency will continue to carry out road-side emissions checks in major towns and cities.

■ The Government will bring forward regulations which will enable local authorities to carry out (with the assistance of police) road-side vehicle emissions enforcement through the use of Fixed Penalties, and also to issue Fixed Penalties to users of stationary vehicles who leave their engines running unnecessarily.

34. The Government also welcomes London Transport's decision to make publicly available the results of smoke checks carried out using Department of Transport equipment.

35. The Government views the continued enhancement of the in-service emission programme as an important element of an overall strategy to sustain the environmental improvements from the UK vehicle fleet. Ongoing effort will be required to improve the effectiveness of enforcement technology which will supplement regular testing. One example is the potential offered by remote emission sensing to target gross polluters. At present, such equipment is incapable of differentiating between a well maintained or poorly maintained vehicle when aimed especially at the car and light duty vehicle fleet. Development work is progressing in the expectation that these shortcomings can be overcome.

■ A 2 year research project is in progress aimed at correlating smoke and particulate emissions from a variety of diesel vehicles operating under simulated road driving conditions with a simple cost effective in-service test.

36. Finally, current technology should make it possible to take emissions tests one step further. On-board diagnostic systems could warn drivers if the emissions control system stops working and prompt them to seek immediate repair.

■ The Government has proposed that the European proposals for post-2000 vehicle standards should include specifications for on-board emissions diagnostic systems for all vehicle types.

37. The Government will ensure that measures to improve maintenance and emissions performance apply to all older vehicles. However, not all vehicles are equally polluting. Lorries, light vans, petrol and diesel cars necessarily have differing emission standards. However, the Government will keep under review the merits of targeted methods of accelerating the reduction of the most polluting vehicles in the fleet.

Traffic Management

38. The purpose of traffic management schemes is to improve the overall quality of life within towns centres. In practice this means not only improving air quality, but also reducing congestion; improving road safety and accessibility; and reducing noise levels. Although the extent of such improvement is difficult to quantify at this stage[2], traffic management schemes can play an important role in improving air quality by: promoting alternative forms of transport such as walking, cycling and park and ride schemes; easing overall congestion; and reducing incentives to drive, for example by making parking less attractive. The Government is issuing guidance on traffic management schemes on the possible ways of reconciling the differing aims of such schemes. Further details of the potential for traffic management are given in Chapter 7.

Planning, Transport and the Need to Travel

39. This Strategy for meeting air quality targets recognises the need for action at national and local level. The effect of integrating air quality objectives into the local planning system and the "package" approach to transport funding, will be longer term. (The broader question of Local Air Quality Management Areas is discussed in more detail in Chapter 7.)

Transport and Development Planning

40. Planning Policy Guidance notes (PPGs) lay down a range of considerations which should be taken into account by local authorities in England when preparing development plans. Of particular relevance to air quality is PPG13, which sets out key considerations for reducing the need to travel and ways of encouraging alternatives to the private car through land use planning. An accompanying guide to better practice for use with PPG13 was published in October 1995, offering a step by step approach to working up new policies and providing examples of good practice. In Scotland, a consultation draft of a National Planning Policy Guideline (NPPG) on Planning and Transport was issued in May 1996. As a further contribution to fostering the intergration of transport and development planning, the Government has recently confirmed that it wishes to go ahead with its proposal to bring the planning of trunk roads in England within the Regional Planning Guidance system. In Wales, guidance is set out in 'Planning Guidance (Wales); Planning Policy' and in a series of Technical Advice Notes (TANs).

[2]Work on this area includes a literature review by the Transport Research Laboratory; this acknowledges inter alia the need for further research

41. The Government recognises that it can be difficult to make the links between transport and land use planning on the one hand, and improving air quality on the other, particularly given the broad range of objectives in each of these areas.

■ A joint central and local government working group has been set up to consider what further advice is necessary to draw together the links between transport and development planning, and air quality.

The package approach to funding

42. As with traffic management, the package approach to funding local transport concerns many aspects of improving quality of life, including air quality. Local authorities in England and Wales are encouraged to take a strategic and multi-modal approach to addressing transport problems by bidding for funding for a package of transport measures rather than for discrete projects. This is known as "package approach" funding. This approach encourages local authorities to assemble bids which set strategic objectives and put forward a coherent package of measures for meeting them ranging across all appropriate modes, looking beyond purely road based solutions. As a general rule, traffic demand management and restraint measures should be included and this, together with proposals to promote and enhance other modes of transport, should aim to achieve modal shift away from the private car. In Scotland one of the main aims of the transport challenge fund is to encourage projects which develop passenger transport systems.

43. Urban areas are the most obvious place in which this kind of balanced strategy can deliver results, but bids are considered from any area on their merits. For example, there have been three successful bids for packages from tourist "honey pots" in rural areas.

44. Action is being taken at national level to enhance the local transport options. For example, the Department of Transport chairs a working group examining ways of making bus travel more attractive. The Government considers that privatisation will breathe new life into the railways to enhance their flexibility and appeal. The Government has brought forward a national strategy for cycling and this has been informed by the innovative schemes in the Government's "Cycle Challenge Competition" in both England and Scotland. The charity SUSTRANS is working in partnership with local authorities to set up a national cycle network aided by a grant of £42.5 million from the Millennium Fund. The Government continues to invest in road and rail, including road by-pass schemes which can do much to ease congestion and improve air quality in town centres.

> *Local Transport Strategies:* The package approach has provided Leeds with encouragement and financial support in combining environmental improvements with an economic regeneration strategy. Within the inner road loop, the environment is largely car free, but a series of alternatives are provided and others are planned to ensure that access to the centre remains viable.

Guidance for Local Authorities

45. A key feature of the Strategy in general, and the Strategy's provisions regarding transport in particular, is the emphasis on local action to deal with local problems. A high degree of co-ordination and cooperation between the different local authorities is needed to ensure that the

Strategy's transport objectives can be met. To this end the Government is issuing guidance on the following subjects:

- guidance on the package approach has been redrafted to ensure that due consideration is given to air quality;

- a circular on air quality and land use planning;

- a circular on traffic management;

- guidance is being issued on reviewing and assessing air quality; and

- guidance on developing local air quality stategies and action plans.

Environmental Responsibility and Individual Action

46. In transport the choices and actions of individual members of the public can have a major impact. The public's role in influencing industrial activity and in practising fuel economy is of great importance, but it is in day-to-day decisions about transport choice, that the individual's determining role in air pollution becomes most significant. Everyone has a part to play.

47. The role of the Government in this area is necessarily limited, but it considers that through the lifetime of this Strategy it should contribute at three levels:

- to help to ensure that accurate information is widely available to members of the public on the impact of vehicle use on air pollution, so that individuals can themselves make informed choices;

- to create a framework in which operators of major vehicle fleets, and in particular public service fleets such as buses and taxis, are encouraged and, where necessary, required to operate on a basis of environmental accountability; and

- in concert with local government, industry and voluntary bodies to help to encourage debate and initiative in this area.

48. The Government sees an important role for voluntary initiatives in this field, such as London First's "Clean Air Charter", which provides advice and an incentive for companies trying to reduce air pollution.

49. To further encourage such initiatives, the Greener Motoring Forum was set up in 1994 to bring together organisations with an interest in road transport. Their remit is to promote practical initiatives that would encourage more environmentally-responsible motoring. A pilot campaign (*Tune Your Car*) was run in five cities during October 1995. Participating garages, dealerships and MOT testing centres offered free, no-obligation, checks on car emissions and where necessary also carried out engine adjustments, free of charge. The results of the campaign indicated that almost every car that exceeds the legal limits for emissions can be put right quickly and cheaply by a simple adjustment. Proposals are being considered for a further campaign.

> *The 1995 Tune your Car Campaign carried out 2400 free emissions tests on petrol-engined cars. On average 33% of cars failed the test. Unsurprisingly, cars with the highest mileage and those which had not been serviced for some time were more likely to fail.*
>
> *The Greener Motoring Forum's new car environmental information scheme will enable car buyers - both individuals and companies - to assess the environmental performance of new cars in terms of factors like fuel efficiency, emissions and noise*

50. The Forum has also produced a free booklet, which explains what motorists can do themselves to improve their car's environmental performance, and is currently developing a new scheme to provide environmental information on new cars.

51. The Transport Green Paper, *Transport: The Way Forward* identified transport intensity as a major issue and the Government is addressing this in three ways:

■ the Standing Advisory Committee on Trunk Road Assessment (SACTRA) has been invited to examine in greater depth the links between traffic growth, transport investment and economic growth;

■ there will be wide-ranging discussions with industry on the scope for reducing transport intensity; and

■ the Government is encouraging and promoting awareness campaigns.

Aircraft, Rail and Shipping

52. Although the main cause of transport related air pollution is due to road transport, it is worth noting that demand for air transport has grown rapidly, particularly over the last 25 years.

53. Emissions from aircraft associated with ground movement and take off and landing cycles (up to 1000 m above the airport) contribute little to overall pollution levels, and are typically much less than the pollution associated with road transport to and from the airport. However, emissions of NO_x by aircraft in the upper atmosphere cause particular concern. Also there remains some uncertainly about the exact impact, the evidence suggests that NO_x is a significant greenhouse gas at altitude. It also has a relationship with ozone.

54. Most civil aircraft fuel is burnt at cruise altitude (8-15 km). NO_x is of particular concern at altitude both as a greenhouse gas and because of its relationship with ozone. Considerable uncertainty remains about the scale of the effects produced by aircraft emissions.

55. Emissions of smoke, hydrocarbons, carbon monoxide and nitrogen oxides are subject to standards agreed through the International Civil Aviation Organisation (ICAO).

■ The Government has played an active role in seeking to introduce tougher standards for NO_x through ICAO, which agreed a 20% tightening of the NO_x standard in 1993. The ICAO Council is considering a recommendation from the Committee on Aviation Environmental Protection for a further tightening of the NO_x standard by 16%. If and when that is ratified, it will be incorporated in UK domestic legislation. Furthermore, the Government supports action through ICAO to remove the exemption of aviation fuel from tax.

56. The UK has been involved with the European project, AERONOX, which has just published its research into modelling aircraft emissions at altitude, and their role in ozone formation and the greenhouse effect. This work is now being taken forward through further projects: POLINAT is measuring emissions in the North Atlantic corridor, and AEROTRACE is providing data on emissions of trace species, including particulates, over the flight cycle.

■ The Government will continue to participate in international discussions on emission standards, taking into account the results of further research and assessment studies on the impact of aviation around airports and at altitude.

57. Over the last ten years, timetable passenger train services increased from 309 million kilometres in 1984 to 350 million kilometres in 1993/94, an increase of approximately 13%. Research undertaken by Railtrack suggests that passenger travel within the UK will show an annual increase of around 2% and that rail's market share will remain broadly stable at around 5% until 2020. Activity in the freight transport market correlates with the level of economic activity. Over the ten year period from 1984 to 1994 (during which there was a real increase in GDP of around 27%), annual freight movements measured in tonne-kilometres increased by around 20%, with rail's market share remaining broadly constant at around 6-7%.

58. The Government wishes to see more freight transferred from the roads to the railways. However, it is expected that rail's share of the market will continue to be affected by strong competition from road haulage. Nevertheless the newly-privatised freight companies are attracting new flows to rail and are operating trains to a number of sites which have not been rail served for a number of years.

59. Rail transport cannot be considered to be emission-free. It is clear that by railway tracks which are used by diesel trains, and near sidings and terminals, pollution from diesel trains can be a significant or even dominant source of pollution. Nevertheless, overall emissions from trains account for less than 1% of the national total. Over the last ten years the percentage of mileage covered by diesel trains, has fallen from 53% to 49%.

60. Emissions from diesel trains should gradually reduce as older stock is replaced by new trains or refurbished with new engines. For example the Class 60 freight locos and Class 165 Turbos have improved fuel consumption control to reduce smoke emissions, together with automatic engine shut-off after a set time. Privatisation of the former British Rail businesses is attracting new interest in rolling stock investment.

61. Developments are not restricted to new locomotives. The InterCity 125 diesel trains are approaching mid-life and franchisees/owners have plans for refurbishment. A number of these units have already undergone refurbishment or renewal of their engines on a trial basis. The resulting fuel economy and greater ease of maintenance makes this approach commercially attractive.

62. Emissions are regulated by the Environmental Protection Act 1990 which requires local government to inspect for statutory nuisances and provides for abatement notices. The Health and Safety Executive also have enforcement duties, where exposure to diesel emissions occur in a workplace such as a station or depot, under the Health and Safety at Work etc. Act 1974. Moreover, it is a condition of train operators' licences under the Railways Act 1993 that an environmental policy is produced, taking account of guidance from the Rail Regulator. This will ensure that environmental standards are maintained, and where necessary improved, after privatisation. Therefore, the Government has no plans for further regulation of railway diesel emissions; existing legislation and enforcement procedures are adequate to deal with any problems which may occur.

- Steps are being taken to reduce emissions in some stations and depots, by providing electricity to stationary trains, thereby reducing the need to idle. Other measures include ensuring that drivers are fully aware of locations which are especially sensitive to smoke emissions and ensuring that maintenance procedures are in place which will identify and rectify faulty engines which produce too much smoke.

63. While shipping is not a major contributor to ambient air pollution problems, it is important that all reasonable steps are taken to minimise the detrimental effects on the environment. The UK is committed to playing an active role in the discussions at the International Maritime Organisation on the development of new regulations to reduce polluting emissions from ships. This will be achieved through a new annex to MARPOL, the International Convention on the Prevention of Pollution from Ships.

Chapter 7: Action at Local Level

Introduction

1. Local authorities, with the support of central Government and the Environment Agencies, will play a major part in the Government's strategy to improve air quality. The ownership by communities of improvements in environmental quality is a key principle running through the Government's Sustainable Development Strategy. Local authorities have the opportunity to build on a long history of local environmental control which, in the case of air quality, was encapsulated in early nuisance legislation. More recent and significant were the Clean Air Acts which did much to reduce the smogs of the 1950's and improve markedly the background levels of smoke and SO_2 in Great Britain. Increasingly, local authorities have exercised their strategic planning and transport functions in the light of air quality considerations. In England and Wales, local authorities administer the system of Local Air Pollution Control (LAPC) under the Environmental Protection Act 1990.

2. The further responsibility for local air quality management is, therefore, a natural progression for local authorities. It is a key part of the Government's overall strategy for delivering sustainable improvements in air quality. The Government believes that the National Air Quality Strategy will deliver a significant improvement. The Government also recognises, however, that it will not be possible to eliminate all potential air quality problems in the most cost-effective way, simply by the use of national policy instruments. There is a significant local dimension to air quality. Because of a combination of local factors and circumstances, which vary markedly from area to area, local hotspots of poor air quality are likely to continue. The Government believes that these are best dealt with at a local level under a new regime of local air quality management.

The Role of Local Authorities

3. Part IV and Schedule 11 of the Environment Act 1995 introduce a new statutory framework for local air quality management. Once fully implemented, local authorities will be required to conduct periodic reviews and assessments of air quality. Where air quality objectives are not likely to be achieved, local authorities must declare Air Quality Management Areas (AQMAs) and make action plans for improvements in air quality in pursuit of national air quality objectives.

4. Local authorities will make a judgement about the content of their action plans and the cost-effectiveness of local action having regard to local circumstances and BATNEEC where appropriate. In some cases, local authorities will not have the necessary powers to bring about the full achievement of air quality objectives, or the achievement of such objectives will not be cost effective. Local authorities are not required to exercise their powers in a way which is not cost-effective. Their principal task is to identify those local air quality hotspots where action at a national level is insufficient to achieve the air quality objectives and to gauge and implement a plan in pursuit of their achievement. Where local action is insufficient to achieve the objectives, it will be for the Government to consider the appropriateness of further action at the national level or the need for supplementary local authority powers.

5. The Environment Act 1995 provides the legal framework for an integrated local authority approach to improvements in air quality but should not be seen as the limit of local authorities' opportunities for improving air quality. While it will quite rightly focus the attention of some local authorities on action in AQMAs, where a concerted effort is most

needed, all local authorities can play a role in improving air quality more generally.

6. Local authorities have existing regulatory responsibilities and will look to employ them to their best effect to secure improvements in air quality. In addition, local authorities will look to reorient their wider planning and transport functions to ensure that air quality considerations are built into their strategic planning processes. Local authorities will be expected to have regard to the Strategy's objectives when preparing land use development plans and when carrying out other duties such as transport planning. Local authorities will also continue their pursuit of sustainable environmental improvement by developing further local policies in the context of the Local Agenda 21 process.

7. This chapter explores further the role of local authorities in improving air quality. The Government considers that the approach from local authorities should be an integrated one, involving all strands of local authority activity which impact on air quality, and underpinned by a series of principles in which local authorities should aim to:

■ secure improvements in the most cost-effective manner, with regard to local environmental needs;

■ seek an appropriate balance between controls on emissions from domestic, industrial and transport sources;

■ avoid unnecessary regulation and promote clarity, consistency and certainty; and

■ draw on a combination and interaction of public, private and voluntary effort.

A Local Air Quality Strategy

8. The Environment Act 1995 gives local authorities new duties and responsibilities which are designed to secure improvements in air quality, particularly at the local level. All local authorities are capable of making a contribution to improvements in air quality both within and outside AQMAs. As part of their commitment to sustainable development they should review the opportunities open to them and develop a local air quality strategy. A local strategy will integrate air quality considerations into a local policy framework and complement other existing local authority strategies.

9. The Government will be issuing further advice to local authorities on the development of local air quality strategies and action plans. The strategic role of local authorities in the management of air quality should draw, however, on the following elements.

10. **A review and assessment of air quality** - it is for local authorities, within the framework of Government guidance, periodically to review and assess air quality within their areas. The Government will continue to operate national monitoring networks for measurement of the major national pollutants. It is important, however, that local monitoring builds on this to give a more detailed picture of local air quality throughout the UK and, in particular, to identify specific, local problem areas.

11. **Sustainable development** - the thrust of the Government's sustainable development Strategy is to embed environmental policies into other, wider policy considerations. Local authorities should embrace this general principle in respect of air quality. Ensuring that air quality is taken into

account, wherever relevant, in local authority decision making and the processes of transport planning and traffic management, land use planning, energy and waste management, enforcement and other local processes would be an important requirement of a local strategy. Drawing together the threads of local policy in this way will be an important consideration of local sustainable development planning and the Local Agenda 21 process.

12. **Cooperation and liaison** - planning positively for the maintenance and improvement of air quality may, in many areas, be most effective over an area wider than an individual local authority's boundaries. In some cases, a single local authority may be the relevant unit for such planning purposes. It is expected, however, that there will be much to gain if local authorities work together over a wider geographical area. In some cases, particularly urban areas, cooperation between local authorities and others will be essential. In any event, all local authorities should liaise and consult fully with neighbouring authorities, and between tiers of local authorities, where appropriate. A framework for appropriate cooperation and liaison among local authorities and with the Environment Agencies, Government, other interested bodies (such as transport agencies) and representatives of the local community will, therefore, form an important part of the local strategy.

13. **Partnerships** - local authorities will also utilise their close relationships with local business and industry and the local community to establish partnerships which can deliver local air quality improvements. The Government believes that a partnership approach offers opportunities for innovative and ground breaking initiatives which otherwise would not be possible or practicable. The Government will, in consultation with local authorities, draw on examples of good practice in this field for inclusion in relevant local authority guidance.

14. **Information and education** - the Government is committed to providing information to the public on ambient air quality and believes that local authorities should play a key role in developing and delivering local information and education. A sense of ownership by local communities remains an essential force for securing local environmental improvement in air as in other media. Local authorities should utilise their involvement with local communities, businesses and the education sector to ensure that information and education reach as wide an audience as possible.

15. Local authorities should consider the particular information requirements associated with AQMAs and be equipped to give appropriate advice to the public during episodes of high pollution. They should also consider the provision of information on air quality where it is relevant to public understanding of local strategies on enforcement, transport and land use planning. Examples of appropriate local authority action will be included in good practice guidance. It is an important principle of developing a local air quality strategy that, as far as is practicable, local communities are involved and kept informed.

16. **A commitment to air quality** - a local authority should, as part of its strategy, consider the impact of its own activities on air quality. The Government is aware that action by local authorities can often act as a catalyst for action by others locally and the record of local authorities in facilitating action in other sectors is a good one. It is important that local

authorities continue to lead local developments in this way. As major institutions and employers, local authorities have the opportunity to make changes to their own working practices to influence local air quality. Consideration, for example, of policies on vehicle use and type would be a useful starting point.

17. **Local air quality management** - the form of a local air quality strategy and the way it is developed will be a matter for local discretion depending on local circumstances and supported, where necessary, by Government guidance. It will be essential, however, that local authorities take action in pursuit of air quality objectives in AQMAs. The Environment Act 1995 sets out, for this purpose, certain statutory duties of local authorities in respect of local air quality management. It also includes reserve powers for the Secretary of State or, in Scotland, SEPA to step in and take action in default where necessary.

18. Local air quality management is a component part of the Government's integrated framework of action to deal with poor air quality. It is particularly important where the National Strategy's objectives are at risk, or where progress in achieving air quality objectives is slow. The Government believes that, in these circumstances, the proper response should be at a local level and the primary responsibility for developing a programme of action should rest with local authorities.

Local Air Quality Management

19. The Government believes that this new framework will:

- support Government action to achieve air quality objectives, particularly where they are at greatest risk;

- promote proportionate and cost-effective action by local authorities;

- focus action on problem areas in appropriate geographical locations;

- be consistent with the UK's obligations under the EC Ambient Air Quality Assessment and Management Directive; and

- enable the Secretaries of State or, in Scotland, SEPA to ensure that appropriate action is taken in those areas where, for whatever reason, local authorities are reluctant to act.

Reviewing and Assessing Air Quality

20. The new duty requires local authorities to review and assess air quality periodically and in a regular and systematic way. The Government is producing guidance to assist local authorities. Although the circumstances of individual local authorities will vary, for many, the review and assessment of air quality will not be an onerous task. The Government believes that, to carry out a review and assessment of air quality appropriate to the needs of each area, many local authorities will be able to rely mainly and primarily on national or local data which is readily available. Guidance will make it clear also that an assessment of whether air quality objectives are likely to be achieved by the year 2005 will not be contingent on the use, by local authorities, of continuous monitoring techniques. The assessment methodology should be consistent with the risk of failure to achieve the air quality objectives in each area thus enabling a variety of techniques to be employed depending on local circumstances.

21. To assist local authorities, the Government is expanding the national database of information from existing monitoring networks and emissions inventories. The Government is also integrating a significant proportion of

local monitoring into a national framework. Further details of the Government's approach to monitoring and assessing air quality are contained in Part II Chapter II.1.

22. As indicated in Chapter 3, the Government does not intend to prescribe air quality standards (as opposed to objectives) in Regulations at this time. The Air Quality Regulations 1997 will prescribe the air quality objectives against which local authorities will assess air quality and determine the need for AQMAs. The air quality standards in this Strategy do not have a timescale attached to them, and represent health-based benchmarks to inform the setting of policy targets - the objectives. It is, therefore, the objectives for the year 2005 which will act as the trigger for local authority action.

23. This does not imply that local authorities should delay the designation of AQMAs where action is clearly necessary. The Government will issue appropriate guidance to local authorities on reviewing and assessing air quality. Where, in the light of this guidance, it is considered that objectives will not be achieved in the appropriate timescale, local authorities should begin the process to create AQMAs as soon as possible. Much of the action local authorities can take in AQMAs will be long term and there will be an inevitable delay, therefore, as programmes of action run their course.

24. The Government is concerned further to ensure that local authorities should not be forced to designate AQMAs to secure the air quality objective in respect of ozone where action by a local authority is unlikely to be cost-effective. Action by local authorities in these circumstances would commit them to unnecessary expenditure and detract them from targeting action to where their involvement brings about the most cost-effective improvements in air quality. For this reason, the ozone objective will not be prescribed in regulations.

25. It will be a matter for local authorities to determine the frequency of reviews and assessments for the purposes of Part IV of the Environment Act. The Government expects, however, that all local authority reviews and assessments will be completed within two years of the new duties on local authorities being fully commenced. In addition, it would expect local authorities to have completed at least one further review and assessment before 2005. The Secretary of State or, in Scotland, SEPA will consider using the reserve powers which the Environment Act provides in the unlikely event that local authorities are not making sufficient progress.

Phased Implementation

26. Following consultation with the local authority associations, the Government announced on 8 February 1996 that a first phase of local authorities would move forward with a review and assessment of air quality in their areas in 1996/97. Some 14 areas were selected to cover a broad range of local circumstances over a wide geographical area. The purpose of this exercise was to:

■ check that initial guidance on reviewing and assessing air quality was suitable for a range of authorities with varying degrees of experience;

■ test the applicability of guidance to areas of differing need, differing complexity and with different sources of pollution; and

■ develop best practice on air quality reviews and assessment and further guidance on air quality management.

27. The first phase also provided an opportunity for the resource

implications for local authorities to be carefully costed as foreshadowed in Air Quality: Meeting the Challenge. As a result of this work, the Government announced that it would be making financial provision in the Revenue Support Grant Settlement for local authorities to fund these new duties. In addition, the Government has created a new Supplementary Credit Approval Scheme in England to assist local authorities with the financial burden of monitoring air quality. In Scotland, capital finance for air quality management is considered under arrangements for single capital allocation.

Air Quality Management Areas (AQMAs)

28. The Environment Act 1995 provides for local authorities to designate as AQMAs those parts of its area where objectives prescribed in the Air Quality Regulations 1997 are unlikely to be achieved. Local authorities will want to consider very carefully whether a designated area should adjoin with the designated area or areas of neighbouring authorities. The Government does not envisage, however, that designated areas will blanket individual or groups of local authorities. The designation of AQMAs will need to be defensible and address the specific local pollution hotspots for which they were designed. Government guidance will give further advice to local authorities on the appropriate designation of AQMAs.

29. Air pollution does not respect local authority boundaries and local authorities acting either individually or collectively in an air quality management grouping will want to plan to address air quality over a much larger area, in conjunction with other agencies, depending on local circumstances. The process of local air quality management must be flexible and responsive to the needs of individual areas. The Government does, however, commend an approach where local authorities work together to plan for improvements over a wider area.

30. Within each AQMA, local authorities will conduct a further assessment of air quality and, within a period of 12 months, prepare a report of that assessment. The Government would expect this report to be reviewed periodically. There will be no fixed period for review, but there will be occasions when a review may be advisable, for example, prior to a major revision of land-use development plans. It will be important for local authorities to appraise development plans and transport plans against these assessments of air quality. In the exercise of their planning, transport and pollution control responsibilities, local authorities should have regard to their assessments of air quality.

Action Plans

31. Where an AQMA has been designated, local authorities must prepare a written action plan in pursuit of air quality objectives in the designated area. This should be completed within two years of designation and include a statement of the timescales in which the local authority proposes to implement the measures outlined in the plan. The boundaries of an AQMA should reflect the location where air quality objectives are unlikely to be achieved. The sphere of influence of an action plan is likely, however, to be much broader and to cover a wider geographic area. This is clearly important in areas where two tiers of local government remain and where actions to improve air quality will be taken at different levels of authority.

32. It will be important that the strategic policies of the various local authorities involved and their local air quality management strategies seek to influence and improve air quality in any designated AQMAs in their

areas. It will also be vital for local authorities to consult with all relevant authorities and interests, especially the Environment Agencies, and, where appropriate, make early contact with the Government Offices in the regions, as they develop their plans. In England and Wales, the Environment Agency and the Local Government Association have developed a protocol setting out a framework for cooperation between local authorities and the Agency. Similarly, in Scotland, SEPA and the Convention of Scottish Local Authorities (COSLA) are developing a Memorandum of Understanding to facilitate co-operation between SEPA and Scottish local authorities.

Local Authority Regulation

33. There will be occasions when local authorities have to resort to enforcement to secure local improvements in air quality. Local authorities remain responsible for a number of statutes designed for this purpose. The Government began a consultation exercise with interested parties on the need for additional local authority regulation in November 1996. The situation will become clearer once local authorities have completed their reviews and assessments of air quality but there may be a need for further regulations if cost effective improvements in air quality cannot be achieved by other means. The Environment Act 1995 provides the Government with wide-ranging powers for this purpose and the situation will be kept under review and reevaluated at the first review of the Strategy in 1999.

Land Use Planning

34. An objective in *Air Quality: Meeting the Challenge* was to ensure that air quality considerations take an appropriate place in the land use planning system. The Government is committed to addressing the following issues:

■ how air quality factors might be taken into account in the development plan process;

■ how planning and environmental protection controls can be more effectively coordinated; and

■ how air quality considerations fit with the broad thrust of current planning policies.

35. Planning Policy Guidance sets out a framework for the relationship between environmental regulation and the operation of the planning system. Whilst existing guidance does not refer specifically to this Strategy or the Environment Act 1995, future revisions will make clear their relevance to the development planning and control process. In the meantime, the Government is consulting on planning and air quality considerations in the preparation of guidance on the linkages between air quality considerations and the planning process. The principal objectives of this guidance are to ensure that:

■ the land use planning system makes an appropriate contribution to the achievement of national air quality standards and objectives; and

■ air quality considerations are properly considered along with other material land use considerations in the planning process.

36. In England, the Government's response to the Transport Debate in England, *Transport the Way Forward,* invited comments on a proposal for integrating more closely the regional land use planning system with the planning of trunk roads. The Government acknowledges the concern of local authorities, and others, that proposals for trunk roads in develop-

ment plans should pay proper regard to regional priorities, local needs and the development pressures they generate. The Government proposes that there should be formal consultation on the trunk road programme during the process of preparing Regional Planning Guidelines. This will enable alternatives such as cost-effective public transport or improved traffic management options to be fully explored at an early stage. The Government will consult with the local authority associations, involving the Highways Agency, to consider whether further measures are needed to improve liaison and coordination. In Scotland, the Government intends to establish a National Transport Forum which will enable it to keep in touch with, and understand the views of, the key operators and policy influences on the Scottish transport scene. The Forum will be chaired at Ministerial level and will comprise key players from local government, transport operators, interested environmental groups and user groups.

Local Air Pollution Control (LAPC)

37. Local authorities in England and Wales share with the Environment Agency the responsibility for controlling emissions from potentially polluting industrial processes[1]. Local Air Pollution Control (LAPC) along with Integrated Pollution Control (IPC), which is exercised by the Environment Agencies, requires the use of Best Available Techniques not Entailing Excessive Costs (BATNEEC) to prevent or minimise, and render harmless, the emissions of certain industrial processes. The new system of local air quality management will complement these existing control mechanisms. The Government has no plans to change the principles which underpin IPC or LAPC, although IPC needs to be reviewed in the light of the requirements of the forthcoming European Integrated Pollution Prevention and Control (IPPC) Directive. BATNEEC remains the cornerstone of the Government's industrial air pollution policies. The systems were designed to deliver progressive improvements in industrial emissions, while taking into account the costs of so doing. It is an important and effective part, therefore, of the Government's Strategy to improve air quality.

38. Currently, LAPC does not control emissions from small combustion plants which have been identified as contributors to localised high levels of SO_2 and NO_x. There is a lower threshold below which authorisation is, at present, not needed. Aggregations of smaller plant, which consequently exceed the threshold, are also exempt. There is a similar threshold for control under IPC but the aggregation of smaller plant does require authorisation. The Government wants to maintain its record of reducing background levels of SO_2 and NO_x. It is anxious, therefore, to remove this anomaly and to simplify the thresholds at which authorisation under LAPC is required. The Government will consult on the options available to it to rationalise the regulation of small combustion plants, including the role which AQMAs might play in triggering new controls.

Smoke Control

39. The Government's smoke control policy has, since 1956, provided an effective system for regulating smoke and ground level SO_2. The limit of designating Smoke Control Areas has now almost been reached but some programmes remain to be completed. In areas such as Belfast in Northern Ireland and the coal mining belt of England, where the use of coal is still widespread, high levels of SO_2 are still recorded. In Northern Ireland, Regulations are due to be implemented which limit the sulphur content of domestic fuel to 2% and strengthen controls on sales of fuels in Smoke Control Areas. If it becomes clear that satisfactory progress is not being

[1]In Scotland, SEPA has responsibility for both IPC and LAPC

made to eliminate the remaining high levels of smoke and SO_2, the Government will review the need for additional policy instruments.

Transport Planning

40. The Government accepts the need for giving increased attention to the environmental impacts of transport. Its Green Papers *Transport - the Way Forward* and *Keeping Scotland Moving* propose that measures to reduce the impact of transport on the environment should be given a high priority, particularly in urban areas where traffic growth poses a particular challenge. Improved transport planning can help in meeting both local air quality and global climate change objectives. There is a balance to be struck, however, between improving air quality and maintaining the accessibility and vitality of urban centres. The Government believes that local authorities are best placed to take a leading role in developing local solutions to local problems. In addition, the Government wants local authorities to take a more strategic approach to local transport planning.

41. To support local authorities' endeavours to achieve these objectives in England and Wales, the Government will continue to provide funding through the package approach which promotes a multi-modal approach to transport planning. Funding sought by local authorities must demonstrate how the measures they wish to implement relate to a wider transport strategy and how that strategy addresses the Government's wider transport objectives. Local authorities in Scotland already have wide discretion to take forward their priority transport projects through the block (capital) allocation system.

42. The Government is committed to working in partnership with local government. It intends to maintain an ongoing dialogue, share expertise and cooperate in research and the dissemination of good practice. The Government also believes that cooperation among neighbouring authorities and between local government tiers is important along with the development of partnerships with business and other interest groups.

Traffic Management

43. The Government believes that many transport decisions to ensure the most effective use of the existing transport infrastructure should be taken at the lowest practicable level. The most appropriate package of local traffic management measures will often be determined at the local authority level, although clearly, in some areas such as London, the need for a wider approach is recognised as important. Local authorities can already deploy a number of traffic management measures to regulate vehicle movement, discourage use of the private car and encourage use of alternative forms of transport, including walking and cycling. These measures are discussed in greater detail in Chapter 6. Local authorities will, as part of their local air quality strategies and their action plans in AQMAs, consider the value of these measures in their local areas. The Government is producing guidance to assist local authorities in evaluating the effectiveness of different traffic management measures and is commissioning further research. The Environment Act 1995 requires local authorities to have regard to the National Air Quality Strategy when exercising functions under the Road Traffic Regulation Act 1984.

44. In certain circumstances, the role of local authorities will need to be enhanced and there may be a case, therefore, for giving local authorities additional permissive powers to manage traffic. To ensure that local authorities have the necessary tools for improving local traffic management, while maintaining local social and economic viability, the Government is discussing with the local authority associations whether additional powers might be appropriate. The case for enabling legislation

to introduce congestion charging, area licensing - where drivers wishing to enter a designated area would need a permit - are considered in Chapter 6.

45. In London, the statutory responsibility for delivering transport and traffic management programmes differs from the rest of the UK. The Government already issues traffic management guidance to London local authorities under the Road Traffic Regulation Act 1991. Following discussions and research, the Government is consulting on revised guidance to reflect recent developments, including the National Air Quality Strategy. The Government wishes to see local authorities in London integrate their approach to air quality considerations with traffic management and parking programmes in the light of the revised guidance. This guidance will complement national guidance on traffic management which is being produced in support of local air quality management. Revised Traffic Management and Parking Guidance calls on local authorities, working together across London, to advise the Secretary of State for Transport of the results of air quality monitoring as far as they relate to transport. It makes clear that any proposals to mitigate emissions from traffic, as part of an air quality action plan, need to be coordinated between affected London Boroughs and with other interested bodies, such as the Highways Agency and the Traffic Director for London.

46. The Government announced in February 1996 that it was to conduct trials in which local authorities would test new powers to check vehicle emissions at the roadside. Proposals will allow suitably qualified local authority staff to carry out vehicle emission tests, and impose, if necessary, a fixed penalty charge. The Government will also consult on a local authority power to penalise drivers of parked vehicles who unnecessarily leave their vehicle engines running. If trials are successful, the Government will move quickly to introduce new regulations for all local authorities.

Urban Environmental Management

47. The principle of local air quality management which underpins action at a national level in support of this Strategy can be applied more widely to address other local environmental issues. In some circumstances, it will be a positive advantage to look at the local environment as a whole. The Government recognises that the problems of localised air quality are likely to be found most commonly in urban areas which also experience other local environmental challenges. The Government is anxious that local strategies to improve air quality do not, as a consequence, exacerbate other local environmental problems. It is possible, however, that local air quality strategies and action plans designed to bring about improvements in air quality in AQMAs will also bring about other localised environmental improvements, noticeably in respect of noise and congestion.

48. There are many factors which influence the vibrancy and vitality of urban environments including, inter alia, the quality of the air, the quality and visual appeal of the built environment, litter, noise, accessibility and safety. The interactions among these factors are often complex and intrinsically local in nature. In some areas, there will be a natural tension between them, which requires a balanced consideration of available solutions. The Government believes that there is merit in ensuring that the impacts on the local urban environment are managed as a package to reflect local needs and circumstances. For example, local authorities are encouraged to consider Community Heating (especially when linked with Combined Heat and Power (CHP)) as a very efficient means of providing affordable heat for local residents, improving the condition of social housing stock, and enabling an overall reduction in emissions through a

more efficient use of energy, even though the beneficial effect on air quality may only be felt at a more diffuse level.

49. Local authorities have a wide range of existing local environmental responsibilities and are, therefore, well placed to secure an appropriate coordinated approach to urban environmental management. They will also be in a position to build on their experience of air quality management and transfer some of its principles to the broader environment. Key among these will be the opportunities for partnerships with other agencies including Government, business and local communities, and the application of local powers such as those being considered for traffic management. The Government wishes to encourage coordinated and focussed local approaches to urban environmental management and will consult with the local authority associations on the opportunities for taking this forward.

Voluntary Measures in the Commercial and Domestic Sectors

50. The previous sections set out a balance of action between central and local government to assess and manage air quality. Important individual and commercial decisions, however, such as those regarding transport also affect the air we breathe, both in the immediate vicinity and more generally. The mode of travel, distance travelled and the level of congestion are important factors, but the ways in which we choose, use and maintain vehicles will also have an impact. Moreover, the places in which we locate our businesses and homes affect their accessibility by public transport or other more sustainable forms of transport.

Commitment from Local Business

51. The traditional view of industrial pollution is that of emissions from a factory chimney. Yet, for the vast majority of modern businesses, the transport demands associated with it can now be the main source of pollution. Long-distance haulage, company car use, urban delivery fleets and the commuting patterns of employees all have an impact on air quality. Many local authorities and businesses are working in partnership to ensure that transport is as efficient as possible and has a minimal impact on the environment.

52. **Company Fleets** - a company vehicle fleet is a significant investment. Effective fleet management can reduce vehicle emissions, promote an image of quality and enhance staff morale. Financial benefits can arise from increased fuel efficiency and reduced mileage. Reduced mileage can be achieved by relatively straightforward steps, such as route planning, mobile communication, and load-sharing with other firms in the same area. Company cars are responsible for more than 35% of total car mileage, yet there are more cost-effective and environmentally beneficial alternatives:

■ a combination of rail and taxi on long journeys may often be less stressful, and increase staff effectiveness once the destination is reached;

■ evidence suggests that personal fuel subsidies result in higher personal mileage and distort decisions about transport options whereas alternative remuneration may be equally attractive;

■ relocation may include the opportunity to reduce car usage and cost.

53. Appropriate driver training can help to reduce poor driving technique, such as stop-start driving, and reduce vehicle emissions. Smoother driving can improve fuel consumption by up to 25% and reduce accidents. Vehicle fleet operators can be encouraged to recognise the environmental and cost benefits of improved driving technique and to aspire to certain

fuel efficiency targets. Vehicle maintenance can also influence emissions performance. Retuning is normally simple and fast and will reduce both excess emissions and fuel consumption. In addition, the use of lower emission vehicles, such as those which run on Compressed Natural Gas or Liquid Petroleum Gas, has considerable environmental advantages in urban fleets of Light Duty and Heavy Goods Vehicles.

54. Further suggestions and details are provided in the booklet *The Company, the Fleet and the Environment* which was produced by the Government in partnership with the Confederation of British Industry. There are also several voluntary partnerships, eg the London First "Green Fleet Charter" which provide detailed guidance for fleet managers.

55. **Green Commuter Plans** - businesses can also influence the ways in which their staff commute to work. Many local authorities are now working together with business to develop plans for reducing car commuting, often through umbrella organisations such as Travelwise. Green Commuter Plans aim to reduce car commuting journeys by a fixed target, in some cases by up to 30%, within an agreed timescale. To be successful, they need to be broadly supported by the staff concerned and by senior management. They can also provide a basis from which business and local authorities discuss and prioritise improvements to local transport systems. Commuter Plans will vary from business to business but aspects which may have a role to play include:

- incentives to buy season tickets, such as easy payment schemes, for public transport or park and ride;

- effective management of car parking;

- information and incentives to encourage car sharing;

- information about public transport; and

- incentives for using bicycles, such as bicycle loans, cycle user groups, and facilities for storing and locking equipment, and showering.

Individual commitment

56. It is difficult to characterise the role of the individual in securing improvements in air quality. Many individuals do, however, make choices about where to live with reference to considerations such as the nearest bus or train routes and whether there is easy access to shops, work and leisure facilities by public transport, walking or cycling. In addition, there are choices which motorists can consider. For example, vehicles sold after 1993 and equipped with a catalytic converter emit far fewer pollutants than earlier models and can reduce emissions by up to 80%. The way in which the vehicle is used and maintained is also important.

57. The Government and local authorities will continue to encourage individual behaviour which reduces the potential to pollute. This primarily involves providing sufficient information for the public to make informed choices. Behaviour to be encouraged includes avoiding making short journeys by car, car-sharing, driving smoothly and at moderate speeds, and ensuring that vehicles are properly maintained.

58. Of course, transport is not the only area in which individual decisions can affect air quality. There are ways in which domestic sources of pollution can be tackled through careful management. Careful energy management in the home, for example, can reduce pollution and save money. In addition, avoiding bonfires when air pollution levels are high and using low-solvent or water-based paints can have implications for local air quality.

Information and Education

59. The National Curriculum for England and Wales, provides a framework for education about the environment to be covered within the subjects of science and geography. Schools are free to build on the compulsory elements in the light of their particular interests and circumstances. A considerable amount of relevant information is already available. In particular, the Government provides public information on a wide range of air quality issues, including health effects and policy, monitoring sites, day to day pollution levels, and individual action.

Assessment of Individual Pollutants

The United Kingdom National Air Quality Strategy

Part II

Part II: Introduction

Introduction

1. This part of the Strategy sets out in more detail the sources and effects of the pollutants considered and the current state of air quality in the UK with respect to the objectives. It also reviews the effects of currently implemented policies and those which it is known will come into force over the next decade. Future air quality in the light of these policies is then assessed against the objectives and any shortfalls or "policy gaps" are identified. Before addressing each pollutant in turn some general issues are discussed below, including a chapter on monitoring, review and assessment.

The Pollutants

2. The pollutants addressed in the Strategy are those which were listed in *Air Quality: Meeting the Challenge* and are those which are being addressed by the Expert Panel on Air Quality Standards (EPAQS). They are benzene, 1,3-butadiene, carbon monoxide, lead, nitrogen dioxide, ozone, particles and sulphur dioxide. EPAQS has not yet published a report on lead but there is already an existing EC Directive (which the Commission is in the process of revising) on this substance so it was considered appropriate to include it in the Strategy. The ninth pollutant on the initial list, polycyclic aromatic hydrocarbons (PAHs) has neither been addressed by EPAQS, nor is there EC legislation on ambient air quality for PAHs. Furthermore, at the present time much less is known about the ambient levels and sources of PAHs than the other pollutants. Accordingly, PAHs will be addressed in more detail in the first review of the Strategy.

3. The pollutants were chosen following a consideration of several factors. Firstly, the pollutants are all known to have adverse effects on human health, at sufficiently high concentrations. Moreover, several of the pollutants can also damage crops, vegetation and ecosystems as well as materials and buildings. In formulating the objectives in the Strategy however, while due regard has been given to these latter effects, the emphasis has been on protecting human health.

4. The second criterion for choosing the pollutants was their widespread occurrence in the UK. As is demonstrated below, in most cases the primary source of ambient levels of the pollutants is the motor vehicle. Where this is not the case (for sulphur dioxide for example) the main sources are similarly widespread small and large combustion plant, or, in the case of ozone, the mechanisms of formation lead to widespread occurrences of elevated levels during certain weather conditions.

5. A third criterion is that there is a reasonable amount known both about ambient levels across the UK and about the major sources of the pollutant, so that coherent abatement strategies can be developed taking due account of costs and benefits.

6. Of course, the list of pollutants arrived at by this process is not exhaustive. Others will be incorporated into the Strategy in the review process, not least those further pollutants which are to be considered by the EC under the Framework Directive, balancing priorities across the Member States. Proposals for these daughter directives are to be brought forward by the end of 1999, with standards and targets therefore not coming into force until the early years of the next century.

Risk Assessments, Costs and Benefits

7. The outcomes of the risk assessments of the individual pollutants is described briefly in the individual chapters which follow. The risk assessments have essentially been carried out by EPAQS in formulating their recommendations. This has involved not only scrutiny of the literature by the Panel itself, but also the advice and assessments of the various Committees and Advisory Groups of the Department of Health, notably the Committee on the Medical Effects of Air Pollutants (COMEAP). Furthermore, the Panel has also had available the risk assessments made by the World Health Organisation (WHO) in revising their air quality guidelines during the past two years.

8. In virtually all of the cases considered, the objectives cannot be met immediately. Accordingly, the Strategy sets timescales over which we will aim to achieve the objectives. In formulating these timescales, each step towards the objective takes account of costs and benefits. However, it must be recognised that many likely costs and benefits, for example: the possible health effects of exposure to very low levels of pollutants; the value of a forest; the visual impacts of development; or global warming, are inherently difficult to quantify, especially in monetary terms. Judgements on each particular case will therefore often need to be made by the appropriate regulator.

9. In this part of the Strategy there is no detailed discussion of formal costs and benefits of particular objectives where policies already in place will deliver the very great proportion of the required improvement. Equally, the Strategy is a broad-ranging document which explicitly provides for relevant review points. Individual measures (part of whose justification may be their contribution towards a long term objective) are, or will be, subject to an analysis of costs and benefits in respect of the specific incremental benefit they offer. Finally, it is likely that in many cases over the next ten years, new and cheaper measures will emerge which could change significantly the costs of meeting some of the objectives.

10. Some indications of these points can be given in some individual cases. For some pollutants, such as benzene, 1,3-butadiene and carbon monoxide, which arise primarily, but not solely, from petrol-engined motor vehicles, current policies are expected to deliver the standards by 2005 and in the majority of areas by 2000.

11. In some cases, involving major sources of pollutants, individual measures have been subjected to detailed assessments of costs and benefits. In most cases there has been, inevitably, more emphasis on the quantification of costs of abatement than of the benefits. Where benefits have been assessed, this has generally been done using measures other than monetary ones. In the case of sulphur dioxide, most of which arises from the electricity supply industry, the recent review of authorisations conducted by HMIP involved detailed scrutiny of costs and benefits - the latter being quantified in terms of exceedences of critical loads and health-related air quality levels. Implementing the outcome of this review should substantially deliver the objectives for this sector.

12. In the important area of vehicle emissions there is a major policy change involving new European proposals for vehicle emission limits and specifications for petrol and diesel fuels. This arises from the so-called Auto-Oil study conducted by the European Commission and Member States with the European vehicle manufacturing and oil industries. Issues of cost-effectiveness were central to the assessments in this study, and

London was one of the key cities addressed in the work. The Commission has recently come forward with its proposals and estimates which suggest that, if agreed, the proposals will go a very long way towards achieving the objectives set in this Strategy for pollutants such as nitrogen dioxide and particles.

Exposure and the Applicability of Objectives

13. In formulating the objectives for air quality policy it is clearly important to address the issue of where the objectives should apply. What is clearly crucial here, given the priority accorded to the protection of human health in the Strategy, is the concept of exposure, which involves consideration of pollutant levels linked to the time over which the pollutant might have an effect. Some pollutants can cause adverse effects over widely differing exposure times, and this will have important consequences for the region of applicability of the objectives, as well as for the monitoring and assessment of air pollutants. Sulphur dioxide, for example, can act over exposure times as short as minutes or even seconds, while at the other extreme, the relevant exposure times for pollutants like lead or benzene are measured in years.

14. Consideration of the locations at which the objectives should apply needs to take into account the likelihood of exposure of the individuals most at risk. Furthermore, these considerations should also have regard to the fact that over a day or longer, a very large fraction - some 80-90% - of the respiratory volume is taken indoors. The approach in the Strategy therefore is to consider that the objectives should apply in non-occupational, near-ground level outdoor locations where a person might reasonably be expected to be exposed over the relevant averaging time of the objective.

15. These considerations ideally demand a knowledge of personal exposures and time activity patterns. At present there is little information available in this area and the Government intends as part of its research programme to characterise the likely personal exposure of the pollutants dealt with here. Nonetheless, the chapters dealing with specific pollutants propose interim approaches.

16. The case for pollutants, such as sulphur dioxide or nitrogen dioxide, where the relevant averaging times for potential effects can be relatively short, is fairly straightforward. Since the sources of these pollutants are widespread, the short term objectives must be considered as applying at any non-occupational near-ground level outdoor location given that exposures over such short averaging times are potentially likely.

17. For longer averaging times, such as 24 hours or a year, the issue is less straightforward. Measurements and assessments at urban background locations probably give a reasonable representation of exposure in these cases, but it is likely that roadsides or other potentially high concentration locations will represent an upper limit to exposures over these periods. At this stage therefore, pending further research, the Government considers that the objectives should apply at such potentially high concentration, non-occupational near-ground level outdoor locations. Consideration of potential exceedences should be carried out in conjunction with data from urban or other appropriate background locations.

Chapter II.1: Monitoring, Review and Assessment

Introduction

1. Effective monitoring and public information systems will be needed for assessing the effectiveness of policies set out in the Strategy, for further developing policies within its framework (both for ambient air quality and for emissions from vehicles and other sources) and to provide the necessary basis for periodic reviews. This chapter summarises how the Government proposes to develop and adapt national monitoring, inventory and modelling capabilities to support the Strategy, and how it proposes to develop its public information arrangements.

Air Quality Assessment: Monitoring, Inventories and Modelling

2. In the present state of understanding of air quality, national monitoring and assessment programmes serve a variety of purposes, including:

■ to help understand air quality problems so that cost effective policies and solutions can be developed;

■ to assess how far objectives are being achieved;

■ to provide public information on current and forecasted air quality;

■ to assist the assessment of personal and environmental exposure to air pollutants;

■ to support local authorities in their review and assessment of air quality;

■ to check compliance with existing European Community Directives.

3. In the context of this Strategy, however, the principal objectives of the Government's development of the UK's monitoring and assessment capacity are:

■ to ensure that the networks are sufficient for, and consistent with, the requirements for measuring compliance with the objectives proposed;

■ to optimise the capacity of the national system through the integration of existing sites and the extension of the networks as a whole;

■ to ensure that an appropriate balance of monitoring, modelling and emission inventory work is maintained.

4. In order to fulfil these objectives, the Department of the Environment funds a number of national monitoring networks, national and urban emission inventories and modelling studies in the UK which are described in more detail below.

Developing Air Quality Monitoring Networks

5. Air quality monitoring in the UK already provides a substantial and accessible source of information for many policy and public information purposes. Many of its features rank with the best in the world, notably its technical sophistication, reliability and the range of pollutants monitored.

6. The overall programmes are organised into a series of networks: three automatic and seven sampler based. As part of its programme to support the operation of the National Strategy, the Government is expanding the network of automatic stations and seeking the integration of national and local government monitoring systems.

National Automatic Monitoring Networks

7. The Department of the Environment funds three national automatic monitoring networks: the Automatic Urban Network, the Automatic Rural Network and the Hydrocarbon Network. The Automatic Urban Network as at January 1997, consists of 54 sites at a variety of locations across the country, including urban background locations representative of the exposure of the population for significant periods of time, and "hotspot" monitoring at the urban kerbside and in the vicinity of industrial sources. Each site monitor some or all of the following pollutants: carbon monoxide; ozone; sulphur dioxide; nitrogen dioxide; and fine particles (as PM_{10}). The Automatic Rural Network consists of 17 sites, monitoring mainly ozone but also sulphur dioxide, nitrogen dioxide and fine particles (as PM_{10}) at rural sites across the country. The Hydrocarbon Network, consisting of 12 sites, monitors 25 volatile organic compounds including benzene, 1,3-butadiene and ozone precursors at urban roadside, urban background and rural locations across the country.

8. The Government is continuing to invest in the expansion of its automatic monitoring networks and established a further 4 multi-pollutant automatic urban sites during 1996, monitoring carbon monoxide, PM_{10}, ozone, nitrogen dioxide and sulphur dioxide. A further 5 multi-pollutant sites will be established during 1997.

9. Many local authorities now also undertake a significant amount of monitoring using real time analyses. This local monitoring can make a valuable contribution to the national effort as part of a harmonised quality assured network. The Government aims progressively to harmonise national and local systems to provide a sound and consistent basis for local air quality management, and has announced its intention to integrate around 35 local authority sites into the national network. Automatic sites from 27 local authorities have already been successfully integrated at Bath, East Birmingham, Bristol, Bury, Exeter, Leamington Spa, South Manchester, Middlesbrough, Oxford, Port Talbot, Rochester, South Somerset, Stockport, Swansea, Thurrock, and in London: Bexley, Brent, Camden, Eltham, Hackney, Haringey (2 sites), Kensington and Chelsea, Sutton (2 sites), Tower Hamlets, and Wandsworth, and the monitoring data are being disseminated together with those measured at the centrally supported sites. Table II.1.1 summarises the present and planned numbers of automatic urban, rural and hydrocarbon monitoring sites in the UK.

Table II.1.1: The UK National Automatic Monitoring Network

Pollutant	No. of existing sites as at January 1997	Additional sites to be established or integrated by end 1997	Total no. of national automatic sites planned by end 1997
Carbon monoxide	41	17	58
PM_{10}	35	14	49
Nitrogen dioxide	56	29	85
Sulphur dioxide	44	21	65
Benzene and 1,3–Butadiene	12	–	12
Ozone	53	21	74

10. The Government does not intend to expand significantly the national automatic Hydrocarbon Network with centrally funded sites. The Government will however keep this under review and will:

■ consider the integration of local authority hydrocarbon sites into the national network;

■ continue to research the use of alternative measurement and assessment methods for these pollutants.

11. The Department of the Environment also monitors a wide range of pollutants in 7 national sampler based non-automatic networks; the Nitrogen Dioxide Diffusion tube Network, the Smoke and Sulphur Dioxide Networks, the Multi-Element and Lead Network, the Toxic Organic Micropollutants (TOMPS) Network, the Acid Deposition Network and the Rural Sulphur Dioxide Network. Two of these networks are operated in co-operation with local authorities.

12. The Government will continue to monitor the rate of decrease in airborne lead in urban areas as the use of unleaded petrol increases. Measurements will also be made, where appropriate, around industrial sources of airborne lead emissions, in consultation with the relevant Agencies.

13. In the light of the rapid expansion of both the automatic and the non-automatic monitoring sites over the last few years the Government will ensure that an audit of the relevance and effectiveness of the national air quality networks will be undertaken by the end of 1997.

Review and Assessment of Air Quality with Respect to Air Quality Objectives

14. The Government will be issuing guidance to local authorities to allow them to undertake consistent and cost effective reviews and assessments of current and future air quality in their locality. This has been considered in more detail in Chapter 7, but in broad terms, the complexity and detail of a review and assessment should be consistent with the risk of failing to achieve air quality objectives by 2005. The three basic tools available for review and assessment of air quality are: monitoring; a consideration of emission sources and the construction of pollutant emission inventories; and dispersion modelling.

15. For practical purposes, a phased approach for the review and assessment of air quality is advocated, in which only those local authorities which have been shown to have a significant risk of exceeding air quality objectives are likely to be involved in the most sophisticated modelling and monitoring techniques and the construction of detailed emission inventories. Even at this stage, less sophisticated monitoring and assessment techniques are expected to have an important role to play.

16. The requirement for each local authority will vary depending on the extent of current modelling or monitoring. Local authorities will need to predict whether a failure to achieve an air quality objective by 2005 is likely. Guidance issued by the Government will assist local authorities to make appropriate predictions.

Monitoring

17. It is important that monitoring systems operate in such a way as to allow assessment of performance against the new air quality objectives. This involves consideration of the number of monitoring stations that are required, the sites at which they should be located, and the monitoring methodology that should be employed. Monitoring regimes have to

reflect the nature and circumstances of individual pollutants. Pollutant-specific issues are therefore addressed in the following chapters, but the general approach is summarised below.

Numbers

18. Where air quality objectives are likely to be exceeded within the relevant period, monitoring sites should be established in sufficient numbers so that, together with other assessment techniques, both the magnitude and geographical extent of exceedences of objectives can be determined. The use of other techniques, such as emission inventory based modelling to assess pollution patterns, is important since it is clearly impossible to do this practicably by monitoring alone.

19. Moreover, with a knowledge of emission patterns it can be shown that fixed air quality monitors can be representative of air quality in other locations. For instance, when pollutants do not show much spatial variability, a single monitoring station can represent a large geographical area. Pollutant concentrations at urban background sites have similar levels within the same city. Ozone sites in rural areas can also be representative of large geographical areas. Other pollutants show a high degree of spatial variability over short distances, for instance carbon monoxide in busy urban centres, but an individual station can be used to represent other similar locations elsewhere.

Location

20. In Chapter 3, the Government set out its air quality standards and objectives. The siting of monitoring stations which are being used to assess achievement or exceedence of the objectives should reflect the exposure of an individual at risk over the appropriate period. The most important locations for monitoring are therefore those at which concentrations are high and a person might reasonably be expected to be exposed for the period of the averaging time specified in the objective.

21. Thus for objectives with short averaging times, such as the 15 minute objective for sulphur dioxide, the sites of greatest potential exposure are likely to be those with highest concentration outside an industrial boundary or within a domestic coal burning residential area. Plumes from power stations can also result in elevated levels in nearby urban areas. For objectives with long averaging times, such as that for benzene, monitoring should focus on measuring long term exposure of the general population and therefore monitoring transient exposure of the population would not be appropriate. However, monitoring of both short and long term standards may be appropriate, for example in locations where housing is next to busy roads or stationary sources.

Monitoring Methods

22. In order to ensure a cost effective monitoring strategy, appropriate techniques should be utilised. In particular, this requires a judgement of the balance between automatic (ie real time) and sampler techniques (ie not real-time). The techniques are complementary, but their systems and aims are different. Automatic sampling can produce continuous, accurate and precise measurements, but generally requires a high degree of organisation and resource. Non-automatic sampling is generally comparatively low cost and therefore is particularly useful for screening, for instance to identify air pollution hotspots, to determine where more sophisticated monitoring may be necessary. Non-automatic passive methods usually provide monthly or weekly averages but are unproved

for some pollutants. Non-automatic active sampling methods can provide daily averages, but are labour intensive and require laboratory analysis.

Emission Inventories and Modelling

23. Emission inventories will be an important aspect of air quality management. They form a means by which local sources can be identified and quantified, an input to models, and a management tool should reductions in sources be required.

24. Detailed emission inventories are most appropriate in the case of larger urban and industrial areas, where specific air quality problems are known to occur, but they will also be relevant to other areas in assessing the scale of any potential problems, or even in demonstrating the absence of problems. Such an inventory will generally require greater detail than is available in the National Atmospheric Emission Inventory which provides information on a 10 by 10 kilometre Ordnance Survey (OS) grid square basis.

25. Recent urban emission inventories for UK cities have been undertaken for London[1], in 1991, and, on behalf of the Department of the Environment, for the West Midlands conurbation[2], in 1996. The London Energy Study which was carried out in 1991 is being updated into a full emission inventory with funding from the European Community. The Department of the Environment is also funding urban emission inventories for a further 10 areas, as outlined below:

Areas | 1-3 | Liverpool and Manchester conurbation
| 4 | Southampton and Portsmouth conurbation
| 5 | Glasgow
| 6 | Bristol and Avonmouth
| 7 | Neath, Port Talbot and Swansea conurbation
| 8 | Sheffield
| 9 | Middlesbrough

The final area is still to be announced.

26. The urban emission inventories are based on a 1 by 1 kilometre OS grid square and take account of point, background and mobile sources. Information is gathered from: domestic and industrial fuel suppliers; traffic and transport models; local authority environmental health departments; EA and SEPA; questionnaires for the sectors of industry not requiring authorisation; assessment of rail, air and water transport; and other inventories. The pollutants covered in the Department of the Environment funded urban inventories include:

Sulphur Dioxide	Black Smoke
Oxides of Nitrogen	Particulate Matter (PM_{10})
Carbon Monoxide	Carbon Dioxide
Total Suspended Particulates	Benzene
Methane	1,3-Butadiene
Non-methane Volatile Organic Compounds	

27. An emission inventory, in conjunction with numerical models, can

[1] *The London Energy Study*, London Research Centre, 1993.

[2] D. Hutchinson and L. Clearley, *West Midlands Atmospheric Emission Inventory*, London Research Centre, 1996.

help in estimating the cost of introducing pollution controls and in identifying the most cost-effective solution for controlling sources.

28. Numerical modelling is a powerful tool which aims to relate the emission of a pollutant to the concentration measured at an air quality monitoring station. The pollutant concentration varies from day to day and from place to place depending on the distribution of pollution sources, meteorology and local topography. The actual levels observed are the result of a balance between a number of competing processes: emissions of compounds into the atmosphere, which may be pollutants in their own right or sources of pollutants such as ozone; their subsequent dispersion and/or chemical transformation into other compounds; and their eventual removal from the atmosphere. Numerical models not only have to incorporate the features of the specific urban environment but also simulate the complete physical and chemical processes involved. Once the model has been developed, it provides a powerful prognostic tool to evaluate different pollution abatement policies.

Information Strategy

29. The dissemination of information on air quality forms a key part of the overall Strategy. One aim of such dissemination is to encourage public debate on environmental decisions by providing clear and up-to-date information. It also allows public scrutiny of the effectiveness of the Government's policies by allowing progress towards the achievement of national air quality objectives to be openly monitored. The presentation of such information can also help to influence patterns of behaviour in more environmentally friendly directions.

30. To this end, the Department of the Environment established an air quality information service in October 1990 through which hourly information, together with an air quality forecast, on a variety of pollutants is widely disseminated through the media. Air pollutants monitored nationally at over 70 automatic sites across the country are currently used to classify air quality as "very good", "good", "poor" or "very poor". Air quality information, together with health advice is provided hourly through CEEFAX (Pages 410-417), TELETEXT (Page 106), an interactive Freephone telephone number (0800-556677) and the Internet (http://www.open.gov.uk/doe/doehome.htm) together with air quality forecasts updated every 12 hours. The existing dissemination system continues to be improved both in terms of the number of pollutants reported and the sites contributing to the network.

31. At local level there is great scope for providing information to the public on the measurements made at locally-operated sites. However, the air quality information obtained locally is sometimes disseminated using locally-derived criteria which differ from those used nationally. There is clearly a need to harmonise the basis on which information is provided to the public. In addition a review of the system is needed to ensure that the national dissemination service reflects the new framework of standards and objectives set out in this document.

32. The Government has published a consultation paper which sets out proposals by which air quality can be described in a consistent manner both locally and nationally. The proposals aim to describe a system which will inform the public as to the current state of air pollution in a manner which clearly indicates any action that they or others might need to take, and reflect the policy framework of the National Strategy.

33. In order to extend and facilitate access to the vast wealth of historical information on air quality that has been amassed over the years, the Government launched the National Air Quality Archive on the Internet in August 1996 (http://www.open.gov.uk/doe/envir/aq/aqinfo.htm). The archive provides a valuable national resource that can be utilised by local authorities, other regulatory bodies, industry, and other interested groups including the public. It includes:

- **automatic monitoring networks** - data and information for ozone, oxides of nitrogen, carbon monoxide, sulphur dioxide, particles as PM_{10} and 25 hydrocarbon species including benzene and 1,3 butadiene;

- **non-automatic monitoring networks** - data on concentrations of lead in air, nitrogen dioxide from the national diffusion tube survey, smoke and sulphur dioxide, toxic organic compounds, and data from the acid deposition network;

- **national atmospheric emission inventory** - disaggregated emissions across the UK on a 10 by 10 kilometre OS grid map for oxides of nitrogen, carbon monoxide, sulphur dioxide, and non methane volatile organic compounds.

Chapter II.2: Research

Introduction

1. A considerable amount of research, funded chiefly by the Department of the Environment, has informed the development of the National Air Quality Strategy, and most of the detailed analyses described in Part II of this document draw upon results of the Department's Air Quality Research Programme. The Government will continue to undertake research on air quality to inform a variety of policy issues, but there will be an increasing emphasis on informing the development and revision of the Strategy. The research programmes will take into consideration the views and research programmes of other Government Departments, notably the Departments of Health, Transport, Trade and Industry and the Welsh and Scottish Offices, as well as EA and SEPA, and the Research Councils, in particular EPSRC and NERC. As is currently the practice, the details of all the individual projects which make up the Air Quality Research Programme of DoE will continue to be published annually.

Future Research

2. Future research in the context of the Strategy will build on the expanded monitoring networks described in the previous chapter. However, important though it is, monitoring of itself cannot provide all the information necessary to manage air quality effectively and efficiently. Research will have to provide the fundamental information required for an understanding of the reasons for the occurrence of elevated levels of pollutants and of the source sector attribution information necessary for successful air quality management. There will therefore be an increased emphasis on analysis and interpretation of monitoring data, and developing an understanding not only of the mechanisms and sources leading to high levels, but also of the reasons behind the observed trends. The continued development of techniques for estimating future trends in emissions and air quality will also be important.

3. Much of this work will be an extension and development of the research currently under way to underpin the guidance which the Government is providing for local authorities in their task of carrying out reviews and assessments of air quality in their areas. The continuing assessment of the adequacy of numerical models and the development of emission inventories will be particularly important, and liaison with local authorities will be necessary in this process.

4. Although the research programme will be developed over time and will be responsive to changing needs, there are already some areas where it is clear that further research will be particularly important. One of these concerns the assessments of costs and benefits. Some work has already been done in this area, but there will continue to be a need to refine the estimates and to incorporate other pollutants. Much progress has been made in formulating dose-response relationships for human health effects in the last few years, but these are still limited to a small number of pollutants. Much more still needs to be understood, and DoE will liaise closely with the Department of Health who are funding a substantial programme of research in this area. It will be particularly important to pay special regard to causal relationships, which play a central role in the enumeration of benefits. Equally, as technology develops and innovation proceeds, abatement costs will change, and the balance between costs and benefits will need to be periodically assessed. This will be a key area in the immediate future.

5. Another area of importance is the subject of personal exposure. This is still a largely uncharted area, although some work is currently under way in the Department of the Environment research programme. As is discussed in subsequent chapters, an understanding of the routes of personal exposure is important in determining where the objectives of the Strategy should apply, and in the estimation of the potential benefits to human health of control policies for air pollution. Of particular importance here is an improved understanding of the relationships between indoor and outdoor air quality and the sources which determine the respective levels in the different environments.

6. The Strategy will need to be flexible in incorporating additional pollutants in due course and the research programme will have to anticipate this. This is discussed further below, but it is already apparent that even with some of the existing pollutants, especially particles and ozone, more research is required.

7. Concerning particles, it is likely that the emphasis in terms of health effects will move to smaller size fractions than PM_{10}. A better understanding of the levels and effects of finer fractions such as $PM_{2.5}$ will therefore be required, as will a quantification of the major sources. Further knowledge is also required of the sources of particles which are present in summertime smog episodes.

8. Ozone presents a difficult problem for air quality management at national and international levels. Research has already indicated the large scale of emission reductions which would be required if the air quality objective is to be achieved. Further research will be needed to determine the extent to which these reductions are achievable given the balances of costs and benefits, both in identifying cost effective control options, and in identifying control strategies which are optimal in a geographic sense across Europe.

9. Research to inform the evolution of the Strategy must also look to the future and in particular to the incorporation of new pollutants. Already it is clear that the first revision will need to incorporate a discussion of PAHs, on which EPAQS has yet to report. Equally there are pollutants such as cadmium, arsenic, nickel and mercury, for which the EU will bring forward daughter directives in the next few years, and the UNECE is bringing forward Protocols on Persistent Organic Pollutants and Heavy Metals. While some information is available on atmospheric levels and sources of these pollutants, further refinement will be necessary in order to assess the extent of any necessary control strategies.

Review Groups

10. It is perhaps appropriate in discussing research in the context of the National Air Quality Strategy, to offer some comments on the system of expert Review Groups operated by the Department of the Environment. This system has played an important part in drawing together the state of scientific knowledge on air quality issues and the Strategy has drawn on the work of QUARG (Quality of Urban Air Review Group) and PORG (Photochemical Oxidants Review Group) in particular. Although the Air Quality Forum, proposed elsewhere in this document, will oversee the development of the Strategy, this body will not be able to provide the detailed scientific assessments of issues in the way that the Review Groups have up until now. Such assessments will continue to be needed, not only to inform the development of the Strategy, but to inform the Government's policies on a wider air quality front.

11. The Government is committed to sound science and the Review Groups provide a valuable assessment of knowledge, drawing on not only the DoE research programme, but also on the wider scientific literature. DoE will continue to convene Review Groups in this way, but reflecting the new aims of the Strategy. Review Groups have worked well when they have had a specific topic to address to inform a particular decision, and this will become the norm for future Groups. The Review Groups' reports have produced wide interest and it is intended that broader interests will be accommodated through an increasing use of peer review, to ensure the widest possible consensus consistent with sound science. The system of Review Groups will evolve with time, but in the immediate future, QUARG and PORG will be combined to address the problem of summertime particles. The role of other existing Groups will be reviewed within this overall framework.

Chapter II.3: Benzene

Introduction and
Sources

1. Benzene is accepted as a human carcinogen. The effect of long-term exposure which is of most concern is leukaemia, and in particular several types of this disease known collectively as non-lymphocytic leukaemia. It has not been possible to demonstrate a level at which there is zero risk of exposure to benzene, and as discussed below, policies to control benzene concentrations in the ambient air therefore adopt a risk management approach, aiming at attaining levels where the risk to health is very small.

2. In the UK the main atmospheric source is the combustion and distribution of petrol, of which it is a minor constituent. Diesel fuel is a relatively small source. The amount of benzene in petrol is regulated to an upper limit of 5% by volume by EC legislation, although currently it comprises on average about 2% by volume in the UK. Motor vehicle exhaust gases contain some of this unburned benzene, but they also contain benzene formed from the combustion of other aromatic components of petrol.

3. Motor vehicles are the most important single source on a national basis, which in 1995 accounted for 67% of total emissions, with 66% of the total arising from petrol vehicles. The emissions from evaporation of petrol and other mobile sources accounted for about 13% of total emission in 1995. Table II.3.1 shows the benzene emission inventory for the UK for 1990-2010.

4. Since the main sources of benzene are motor vehicles, and primary emissions are dispersed and diluted, benzene is a local rather than a transboundary pollutant.

**Table II.3.1:
Emissions of benzene
in the UK, 1990-2010**

| Source | Emissions (tonnes)* | | | | | Percentage of Total in 1995** |
	1990	1995	2000	2005	2010	
Petrol exhaust	30,500	23,000	11,930	7,350	5,680	66
Diesel exhaust	340	440	560	610	680	1
Petrol evaporation	2,820	1,890	2,060	1,270	1,030	5
Other Mobiles	2,540	2,500	2,040	1,730	1,730	7
Stationary Combustion	3,100	2,400	2,230	2,140	2,180	7
Extraction and Distribution of Fossil Fuels	1,480	1,290	1,000	950	910	4
Iron and steel manufacture	600	580	320	320	320	2
Industrial Processes	2,970	2,580	1,440	1,410	1,400	7
Landfill	170	160	160	160	160	1
Total	**44,520**	**34,840**	**21,740**	**15,940**	**14,090**	**100**

* Figures rounded to nearest 10 tonnes
** Figures rounded to nearest 1%

Chapter II.3: Benzene

Health Effects:
**Standards and
Guidelines**

5. It was noted in the Introduction to this Chapter that it is not possible to demonstrate a completely "safe" level of benzene at which there is zero risk. The World Health Organisation have accordingly published an air quality guideline, but have quoted a so-called unit risk factor, which is the risk (in the case of benzene of developing leukemia) associated with lifetime exposure to unit concentration. This approach relies on being able to extrapolate the effects of what are generally much higher doses involved in animal experiments or epidemiological studies based upon occupationally exposed humans to the low doses found in ambient air. EPAQS, taking advice from the Department of Health Committee on Carcinogenicity, did not feel that such extrapolations could be carried out with confidence and approached the problem differently. EPAQS considered the medical evidence for carcinogenic effects from benzene and derived a recommended standard, 5 ppb as a running annual mean, which would represent an exceedingly small risk to health. Further, acting on the advice of the Committee on Carcinogenicity that exposure to benzene should be kept as low as practicable, EPAQS recommended a target of 1 ppb, also as a running annual mean[1].

*Current Air
Quality*

6. Spot measurements of benzene concentrations in ambient air have been made in the UK since the 1970s. The first long term time series measurements were made in rural locations. These early measurements in rural areas indicated an annual average concentration less than 1 ppb (0.34-0.81 ppb).

7. Continuous hourly measurements in urban areas began in July 1991 at a roadside site in Exhibition Road, London and continued for a year as part of the DoE Research Programme. The annual mean benzene concentration during this period was 4.1 ppb. These measurements confirmed that motor vehicle traffic accounted for the greater part of the benzene concentrations found at that roadside location.

8. The automatic Hydrocarbon Network measurements provides hourly data on benzene. Currently sites are operational in London (Bloomsbury and Eltham), Birmingham, Cardiff, Edinburgh, Belfast, Middlesbrough, Bristol, Harwell, Southampton, Liverpool and Leeds. All except two of these site locations are described as urban background, that is they are in central urban areas, away from the immediate influence of roads, being at least 20 metres away from the nearest road. One site (London Bloomsbury) samples about 4 metres from the kerbside of a moderately busy road. There is one rural site at Harwell in Oxfordshire.

9. The currently available measurements are summarised in Table II.3.2. In general terms, given the locations of the sites, the concentrations in Table II.3.2 are considered to be typical of the urban exposure levels of a large section of the general population. Annual mean concentrations are comparable. The London Bloomsbury site shows the highest concentrations on average, Edinburgh is the site with consistently lowest urban concentrations. The benzene concentrations are lower by a factor of 2 to 3 compared to those reported for Exhibition Road in London during 1991-92. Recent comparisons of relatively sparse benzene data for London collected since the early 1970s suggest a substantial decline in

---

[1] Expert Panel on Air Quality Standards, *Benzene.* London, HMSO, 1994.

benzene concentrations over the last twenty years. Continuous hourly measurements have yet to be made in the UK at the kerbside of the most heavily trafficked roads. Nonetheless, the benzene measurements that already exist from Exhibition Road can be used to estimate kerbside benzene concentrations using the Cromwell Road kerbside NO_x and carbon monoxide concentrations. On this basis it is estimated that mean benzene concentrations at the kerbside adjacent to heavily trafficked roads may lie in the range 9.6-11.6 ppb.

Table II.3.2: Annual benzene concentrations at UK sites (ppb)

Site	Year	Annual Mean	Max RAM*	% Data Capture
Belfast	1994	1.10	1.73	96.8
	1995	0.90	1.10	95.1
Birmingham	1994	1.03	1.71	91.9
	1995	1.04	1.10	95.5
Bristol	1994	1.06	1.62	62.7
	1995	1.22	1.23	93.3
Cardiff	1994	1.50	1.95	94.4
	1995	1.22	1.51	86.3
Edinburgh	1994	0.70	1.01	91.0
	1995	0.74	0.74	89.8
Harwell (rural)	1994	-	-	-
	1995	0.38	-	73.9
Leeds	1994	-	-	-
	1995	0.98	0.98	94.5
London Eltham	1994	1.14	1.24	83.4
	1995	1.02	1.16	91.3
London Bloomsbury	1994	1.78	2.26	90.8
	1995	1.70	1.78	91.2
Middlesbrough	1994	1.29	1.37	64.5
	1995	1.08	1.29	95.7

* Maximum running annual mean that occurred within the calendar year.

10. A significant amount of benzene monitoring using diffusion tubes has been carried out in the London area. Diffusion tube measurements are useful for area-wide surveys of benzene if sampling periods can be kept short enough and if ambient concentrations are high enough. They can be problematic particularly with extended sampling periods because of the reversible uptake of benzene on the adsorption medium deployed. Mean benzene concentrations of up to 37 ppb have been reported at roadside sites in close proximity to the kerbside. Typical mean benzene concentrations from diffusion tube measurements at roadside sites within 20 metres of busy roads appear to cover the range 3-14 ppb. Such levels are consistent with the mean benzene concentrations estimated for Cromwell Road in the previous paragraph.

11. Benzene emissions are reported from industrial sources, in addition to motor vehicles, and so there are questions concerning this contribution to urban exposure levels. The availability of two years' data from the

continuous hydrocarbon monitoring site at Longlands College, Middlesbrough presents an opportunity to examine benzene levels in an urban background location which is potentially influenced by localised, industrial emissions. The hourly data for benzene shows the clear influence of sporadic peaks of benzene superimposed upon a steady baseline in a manner not found at most of the other automatic monitoring sites. These peaks have been associated with industrial sources and the baseline with motor vehicle traffic. Despite there being significant peak hourly mean benzene concentrations of up to 55 ppb, annual mean concentrations are barely different from those reported for other urban background sites. Motor vehicle benzene emissions still appear to make a dominant contribution to annual mean ground level benzene concentrations in urban areas with significant industrial benzene sources. Account should be taken of exposure of populations living downwind of such local sources. It is likely that their exposure will be greater than that suggested by data collected at a distant monitoring site which may only register local point source emissions on certain wind directions.

12. In considering current air quality in relation to the standards recommended by EPAQS, it is first appropriate to consider the recommended standard of 5 ppb running annual mean. This is likely to be exceeded currently in urban locations at the roadside (or within about 20 metres) of heavily-trafficked roads. At the busiest of roads, levels may be as high as double the standard recommended by EPAQS and in some localised cases even higher, possibly as much as eight times the standard.

13. An estimate of the concentrations of benzene across the UK can be made by using relationships between mean benzene and nitrogen dioxide and a map of nitrogen dioxide concentrations across the UK obtained from measurements with diffusion tubes at some 400 sites. In this way concentrations can be estimated for each 5 by 5 kilometre OS grid square in the UK and are shown in Figure II.3.1.

14. Exceedences of the longer-term target standard of 1 ppb annual mean recommended by EPAQS are necessarily more widespread. The mean concentration at all but two of the Department of the Environment urban background monitoring sites exceeds the EPAQS target value. Almost all locations within 100 metres or so of a busy road are likely to experience concentrations which approach or exceed this value.

15. The extent of the exceedences of the 1 ppb level can be assessed in the same way as the 5 ppb recommended standard, and this is shown in Figure II.3.2. The exceedences are distributed throughout the major urban areas of the UK. This conflicts with the available measurements in some areas which to date show mean values less than 1 ppb. This suggests either that the method of mapping used here may overestimate concentrations or that there are some locations where the EPAQS target value is being exceeded which are not being picked up by the monitoring sites.

The Strategy 16. The effects of current policies have been used to estimate the reductions in emissions likely to occur in future years. The significant reductions in emissions from motor vehicles obtained through the implementation of vehicle exhaust emission limits are offset to an extent through the forecast increase in traffic activity over the next decade or so.

17. In making the estimates of future air quality here, the National Road

Traffic Forecasts published by the Department of Transport have been used, together with the assumption that the following policies are implemented:

- 'EURO I' vehicle emissions Directives which set limits for the emissions of nitrogen oxides, hydrocarbons, carbon monoxide and particles (diesel vehicles only). In practice EURO I required the fitting of catalytic converters to all new petrol cars;

- Directive 93/441/EEC, passenger cars (from 31/12/1992);

- Directive 93/59/EC, light vans (from 1/10/1994);

- Directive 91/542/EEC, heavy duty vehicles (from 1/10/1993);

- it has further been assumed that diesel car sales would level off at 20% of all car sales from 1994;

- for petrol evaporative emissions, Petrol Vapour Recovery Directive 94/63/EEC (PVR) controls effective from 1996 on new installations and 1999 on existing facilities;

- it is not clear at present what effects the 1996 Stage II exhaust emission limits will have on benzene emissions, so that the effect of this Directive has not been included.

18. On the basis of these assumptions, future benzene emissions can be estimated and these are shown in Table II.3.1 for the years 2000, 2005 and 2010, together with current emissions for comparison. On the basis of these projections, total benzene emissions are expected to decline by almost 40% by the year 2000, over 50% by 2005 and by about 60% by 2010, on a 1995 base.

19. Assuming that these percentage reductions in benzene emissions are achieved evenly across the entire UK, then it is likely that by the year 2000 there will be no more exceedences of the 5 ppb levels recommendation at urban background locations and at most roadside locations next to heavily-trafficked roads. There may still however, be some roadside environments which, because of their close proximity to the most heavily-trafficked roads, will continue to exceed the 5 ppb level, although these exceedences are likely to be small. These environments should in turn comply with the 5 ppb level by 2005.

20. The impact of the projected decrease in benzene emissions from motor vehicle traffic by the year 2000, will be to reduce substantially the urban background concentrations of benzene. It is estimated that the number of 5 by 5 kilometre OS grid squares in which the target 1 ppb running annual mean level is likely to be exceeded in the year 2000 will have been reduced significantly. This dramatic reduction in the area exceeding the target is expected to have resulted from the reductions in benzene concentrations in suburban locations where the target is currently exceeded by only a small margin. Reductions in urban background benzene concentrations are unlikely to be large enough to bring benzene concentrations below the target in the majority of urban population centres.

21. On present expectations, benzene emissions from road traffic are anticipated to decrease further between the years 2000 and 2010. The number of 5 by 5 kikometres OS grid squares in which the target benzene level of 1 ppb is likely to be exceeded in the year 2010 should have been reduced dramatically and areas of exceedence should have been

eliminated in Scotland, Wales and Northern Ireland. However, most of the large population centres will still have some areas in which it is expected that the target benzene level of 1 ppb is exceeded.

Conclusions 22. Benzene is a genotoxic carcinogen so that no absolutely safe level can be specified for ambient concentrations. Therefore the intention must be to reduce ambient levels such that they represent an exceedingly small risk to human health.

23. The Government notes EPAQS' judgement that at levels around 5 ppb, the risks to public health are exceedingly small and the Government accepts this value as the standard as a running annual mean. It is clear from the foregoing that current policies, if rigorously enforced, should eliminate exceedences of the standard by 2005. The Government has decided therefore to adopt as an objective the level of benzene of 5 ppb as a running annual mean, to be achieved by 2005.

24. The Government has decided that the objective should apply in the following non-occupational, near-ground level outdoor locations: background locations; roadside locations; and other areas of elevated benzene concentration where a person might reasonably be expected to be exposed (e.g. in the vicinity of housing, schools, or hospitals etc) over the averaging time of the objective. The latter areas are likely to represent an upper limit to exposures over a period of a year, but recent research, funded by the Department of the Environment, in the Avon Longitudinal Study of Pregnancy and Childhood Study, showed that outdoor benzene levels were a significant determinant of indoor levels. Consideration of potential exceedences of the benzene objective at potentially high concentration locations should be carried out in conjunction with data from urban or other appropriate background locations.

25. Policies already in hand should, furthermore, reduce dramatically the exceedence of the longer term target level of 1 ppb recommended by EPAQS. Although there may still, on current projections, remain a small number of areas where this target level is exceeded, it will be treated as an indicative aim of policy, with the intention of reducing levels as far as practicable below this limit by 2005 and beyond.

26. Further policies to maintain this level of air quality over timescales appropriate to considerations of sustainability will need to address the main sources of benzene remaining beyond 2005. These are:

■ petrol-engined vehicle exhausts, due particularly to deterioration of catalyst performance with age;

■ petrol refining and distribution, due mainly to the fuel losses not controlled by small canisters, and uncontrolled emissions from petrol station forecourts without vapour recovery systems.

27. The complete elimination of exceedences of the 1 ppb target level would therefore require policies to deal with emissions from these sources. Some of these policies would also have beneficial effects on other pollutants as well as benzene, so that assessments of costs and benefits would need to go beyond the benefits arising from lower benzene levels alone. Additional policy measures to deal with benzene and other emissions might include:

■ further reductions in vehicle emission limits post 2000;

■ reductions in the sulphur content of petrol to reduce the deterioration in catalyst performance with age;

■ reductions in aromatic and benzene content of petrol to minimise the impact of the imperfect control performance of catalyst and evaporative emission controls;

■ controls on petrol vapour recovery from refuelling vehicles at service stations;

■ stricter emission limits for oil refineries and stationary combustion sources.

28. The first three of these points are being considered in the formulation of vehicle emission standards and fuel quality within the EU. The Government will press for significant improvements over the 1996 standards in these discussions.

Chapter II.3: Benzene

**Figures II.3.1:
Estimated urban
background annual
mean benzene
concentrations 1994 (ppb)**

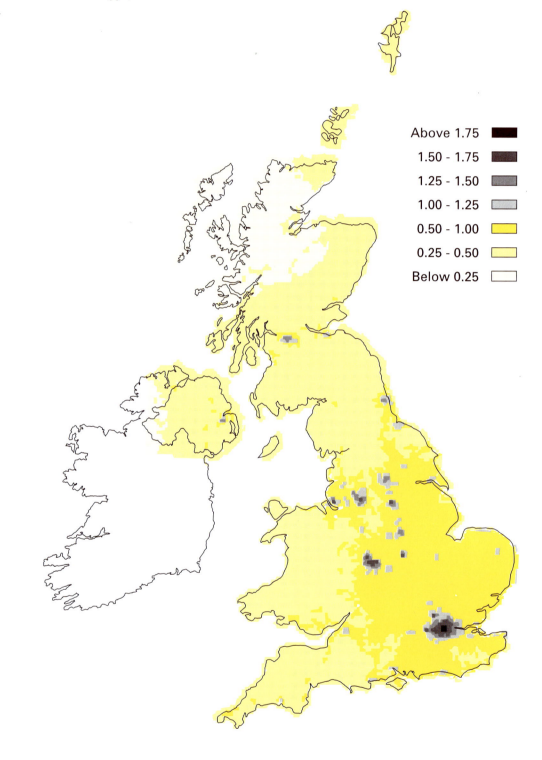

Above 1.75

1.50 - 1.75

1.25 - 1.50

1.00 - 1.25

0.50 - 1.00

0.25 - 0.50

Below 0.25

**Figures II.3.2:
Exceedences of 1 ppb
running annual mean for
benzene 1994**

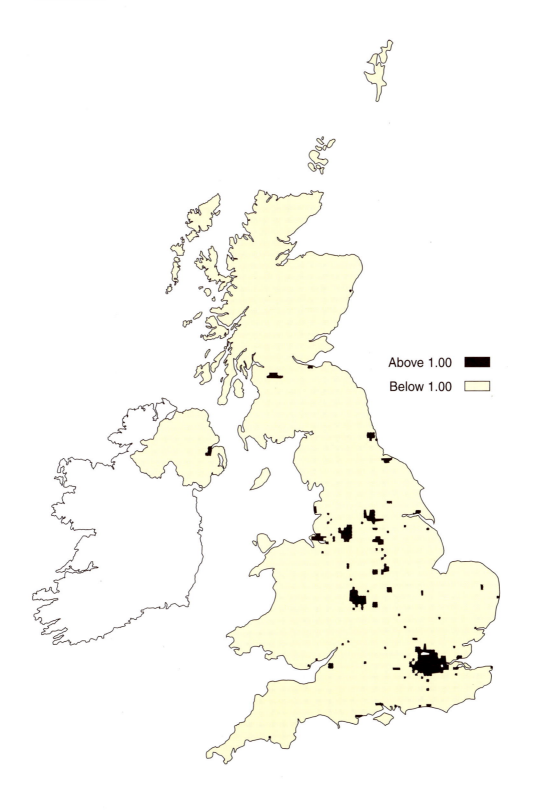

Above 1.00 ▪
Below 1.00 ▫

Chapter II.4: 1,3-Butadiene

Introduction and Sources

1. 1,3-Butadiene is a gas at normal temperatures and pressures and trace amounts are present in the atmosphere, deriving mainly from the combustion of petrol and of other materials. 1,3-Butadiene is used in industry, mainly in the production of synthetic rubber for tyres. It is thus a chemical to which workers have been exposed, and there is evidence that such groups of workers have had a slightly higher than expected risk of cancers to the bone marrow, lymphomas and leukaemia. Laboratory studies have shown that 1,3-butadiene causes a variety of cancers in rodents and damages the genetic structures of the cell. It is thus a genotoxic carcinogen and, in theory, it is not possible to determine an absolutely safe level for human exposure.

2. Although neither petrol nor diesel fuel contain 1,3-butadiene it is formed in the combustion process from olefines in the fuel. The proportions of these olefines in petrol have been increasing over the last decade and it is likely that the amounts of 1,3-butadiene emitted into the atmosphere from road traffic have also been rising. 1,3-Butadiene is also an important industrial chemical. It is handled in bulk at a small number of industrial locations in the UK. Other than in the vicinity of such locations, the dominant source of 1,3-butadiene in the UK atmosphere is the motor vehicle.

3. An emission inventory for 1,3-butadiene is shown in Table II.4.1 from which it can be seen that in 1995 petrol vehicles contributed 67% of national emissions. Emissions of 1,3-butadiene from industrial sources, either manufacturing or using the compound, contribute some 13% at a national level. Since this latter emission will occur in a relatively small number of locations, levels of 1,3-butadiene in the large majority of urban areas will be dominated by emissions from motor vehicles. However in those areas in the vicinity of industrial emissions, the latter will make a proportionally greater contribution.

Health Effects: Standards and Guidelines

4. EPAQS accepted that 1,3-butadiene is a genotoxic carcinogen and that, as in the case of benzene, no absolutely safe level can be defined. The Panel nevertheless believed that a standard could be set at which for practical purposes the risks are exceedingly small and unlikely to be detectable by any practicable method. The Panel recommended[1] a standard of 1 ppb as a running annual mean and, recognising the uncertainties in the data on effects on humans and the limited amount of monitoring data at present, recommended a review in 5 years.

Current Air Quality

5. 1,3-Butadiene is a reactive gas whose atmospheric measurement is difficult and requires sophisticated instrumentation. Measurements of 1,3-butadiene were first made on a regular basis in 1990 by the University of East Anglia, at rural sites in north Norfolk and on the hill-top site at Great Dun Fell in Cumbria. The first regular urban measurements were made at the Imperial College roadside site at Exhibition Road in London from 1991-92.

6. In 1992 the Department of the Environment established a network of sophisticated automatic gas chromatographs which are equipped with telemetric data collection. Through this network, data on 1,3-butadiene

---

[1] Expert Panel on Air Quality Standards, 1,3-Butadiene, London, HMSO, 1994.

(and 24 other volatile organic compounds including benzene) are made available to the public on a daily basis. At present a total of 12 sites across the UK measure 1,3-butadiene on an hourly basis. A summary of available data is presented in Table II.4.2. Annual average concentrations in urban areas range from 0.10-0.38 ppb, with rural concentrations somewhat lower.

7. As was the case with benzene, there are at present no data available for 1,3-butadiene at the kerbside of heavily trafficked roads. However, using co-located measurements of carbon monoxide, NO_x and 1,3-butadiene at Exhibition Road together with carbon monoxide and NO_x at the heavily trafficked Cromwell Road site, 1,3-butadiene concentrations can be estimated at 1.7-2.7 ppb at the latter site.

8. Mean concentrations of 1,3-butadiene have been estimated for each 5 by 5 kilometre OS grid square across the UK. This was done using the observed 1,3-butadiene and nitrogen dioxide concentrations and the 1991 diffusion tube survey of six monthly mean nitrogen dioxide concentrations carried out by the former Warren Spring Laboratory. The results of these are shown in Figure II.4.1. A small number of squares in the London area are expected to have mean 1,3-butadiene concentrations which exceed the value recommended by EPAQS.

9. On the basis of the available data therefore, the EPAQS standard is unlikely to be exceeded at the present time, except at a small number of urban background and heavily trafficked roadside locations.

The Strategy   10. The policy measures of relevance to the control of 1,3-butadiene concentrations are the vehicle emission limits for petrol and diesel vehicles to control mobile source emissions and the Environmental Protection Act 1990 (EPA 90) through which the EA can regulate emissions from the major industrial sources of 1,3-butadiene.

11. In the great majority of towns and cities, motor vehicles are the dominant source of 1,3-butadiene. A major industrial source is located near the Middlesbrough measurement site and recent work has shown that, while fairly large short term peak concentrations can be attributed to industrial releases, these make only a small contribution to the long term average level of 1,3-butadiene which is dominated by traffic sources. Nonetheless, releases into the atmosphere of potentially genotoxic carcinogens are clearly of concern and, where appropriate, measures under the EPA 90 will be taken to minimize such releases.

12. The effect of the introduction of three-way catalysts into the vehicle fleet is anticipated to make a significant reduction in urban levels of 1,3-butadiene. Measurements of emissions from in-service vehicles on realistic road drives have shown that three-way catalyst equipped vehicles emit less than 10% of the 1,3-butadiene of a non-catalyst car, even on cold-start drives.

13. United Kingdom 1,3-butadiene emissions from petrol cars were estimated to be about 5500 tonnes in 1992 on the basis of the emissions in central London required to account for the observed concentrations. Current UK 1,3-butadiene emissions from petrol cars are estimated to be about 6400 tonnes per annum. Emissions in the years 2000 and 2010 have been estimated at 2500 and 1500 tonnes per annum respectively, which represents a reduction of 61% and 76% from current levels. On this

basis, petrol-engined motor vehicle 1,3-butadiene emissions are expected to decline by about 55% by the year 2000 on their 1992 values and by 73% by the year 2010.

14. Assuming that these percentage reductions in 1,3-butadiene emissions are achieved evenly across the entire UK, then it is likely that by the year 2000 there will be no more exceedences of the level recommended by EPAQS in urban background locations and in most roadside locations next to heavily trafficked roads. There may well, however, be some roadside environments which, because of their close proximity to the most heavily trafficked roads, will continue to exceed the EPAQS level for some time to come. However, by 2005 the emission reductions currently foreseen are likely to be such that the 1 ppb value is attained in virtually all locations.

Conclusions

15. 1,3-Butadiene is, like benzene, a genotoxic carcinogen, so again no absolutely safe level can be specified for ambient concentrations. Therefore the intention must be to reduce ambient levels such that they represent an exceedingly small risk to human health.

16. EPAQS recommended the value of 1 ppb as a running annual mean, which the Government accepts as the standard for 1,3-butadiene. The air quality objective will be to achieve this standard by 2005. The EPAQS recommended standard will be reviewed by 1999 and, depending on the outcome of that review, further assessments will be made, in conjunction with local authorities where appropriate, of the likelihood of any local exceedences.

17. The Government has decided that the objective should apply both in the following non-occupational near-ground level outdoor locations: background locations; roadside locations; and other areas of elevated 1,3-butadiene concentrations where a person might reasonably be expected to be exposed (e.g. in the vicinity of housing, schools or hospitals etc) over the averaging time of the objective. Consideration of potential exceedences of the 1,3-butadiene objective at potentially high concentration locations should be carried out in conjunction with data from urban or other appropriate background locations. As with benzene this approach will be reviewed in the light of further research into patterns of personal exposure.

Table II.4.1: Emissions of 1,3-butadiene in the UK, 1995

Source	Emissions (tonnes)★	Percentage of total★★ in 1995
Petrol Vehicles	6,390	67
Diesel Vehicles	1,030	11
Use of Butadiene as Feedstock	610	6
Butadiene Manufacture	630	7
Gas Leakage	400	4
Landfill	510	5
Total	9,570	100

★   Figures rounded to nearest 10 tonnes
★★   Figures rounded to nearest 1%

**Table II.4.2:
Annual 1,3-butadiene
concentrations at UK
sites**

Site	Year	Annual Mean	Max RAM*	%Data Capture
Belfast	1994	0.18	0.20	95.3
	1995	0.17	0.18	94.0
Birmingham	1994	0.22	0.34	91.4
	1995	0.24	0.25	97.1
Bristol	1994	0.22	0.53	41.8
	1995	0.24	0.24	82.6
Cardiff	1994	0.27	0.41	94.1
	1995	0.23	0.27	83.8
Edinburgh	1994	0.11	0.40	79.0
	1995	0.13	0.13	84.8
Harwell (rural)	1994	–	–	–
	1995	0.18	–	16.0
Leeds	1994	–	–	–
	1995	0.21	0.21	94.3
London Eltham	1994	0.24	0.25	71.9
	1995	0.24	0.26	87.2
London Bloomsbury	1994	0.38	0.45	90.3
	1995	0.36	0.38	90.0
Middlesbrough	1994	0.26	0.28	63.3
	1995	0.27	0.31	91.5

* Maximum running annual mean that occurred within the calendar year.

Figures II.4.1:
Estimated annual mean
1,3-butadiene
concentrations
1994 (ppb)

Above 0.4
0.3 - 0.4
0.2 - 0.3
0.1 - 0.2
Below 0.1

Chapter II.5: Carbon Monoxide

Introduction and Sources

1. Carbon monoxide (CO) is a gas formed by the incomplete combustion of carbon containing fuels. In general, the more efficient the combustion process, the lower the carbon monoxide emission. At very high levels, prolonged exposure to carbon monoxide can result in death. At lower levels, the reduction in the oxygen-carrying capacity of the blood may increase the risk of heart problems in predisposed individuals. Although the major concern over carbon monoxide in the UK relates to very high indoor concentrations arising from faulty combustion appliances causing fatalities, there are also potentially adverse effects on health from high levels in the outdoor environment.

2. The main source of carbon monoxide in the UK is road transport which currently accounts for almost 75% of the emissions of some 4.1 million tonnes per year (see Table II.5.1). Of this road transport emission, the predominant source is petrol vehicles which account for 71% of the UK total and 95% of the road transport emissions. Emissions of carbon monoxide in the UK have increased significantly from 6.50 million tonnes in 1970 to 7.38 million tonnes in 1990, an increase of 13%, shown in Figure II.5.1. Emissions have, however, been decreasing since 1990. Emissions of carbon monoxide from diesel vehicles are relatively small and currently contribute 4% of the national total.

3. Since the main source of carbon monoxide is motor traffic, concentrations are highest near to heavily trafficked roads. Concentrations fall fairly rapidly with distance away from roads so that carbon monoxide

Table II.5.1: Emissions of carbon monoxide in the UK, 1990-1995

Source	Emissions (kilotonnes)*						% of total in 1995**
	1990	1991	1992	1993	1994	1995	
Road Transport:							
Petrol	5364	5325	5029	4644	4278	3917	71
Diesel	181	188	182	185	194	195	4
Power Stations	294	297	300	259	240	232	4
Domestic	434	458	423	444	395	341	6
Commercial/ Public Service	8	8	7	6	5	5	<1
Refineries	5	6	6	6	6	6	<1
Other Industry	990	985	940	746	741	667	12
Offshore oil & gas	33	35	36	44	47	48	1
Railways	12	12	12	12	11	11	<1
Aircraft	1	11	11	12	12	13	<1
Shipping	17	18	18	18	17	17	<1
Military	8	7	7	6	6	6	<1
Agriculture	20	20	20	20	20	19	<1
Total	**7377**	**7370**	**6991**	**6402**	**5973**	**5478**	**100**

* Figures rounded to the nearest kilotonnes

** Figures rounded to nearest 1%

Chapter II.5: Carbon Monoxide

Figure II.5.1: Emissions of carbon monoxide

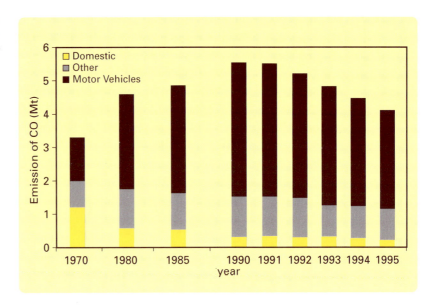

is a local, rather than a transboundary pollutant. Carbon monoxide is also an indirect greenhouse gas which influences atmospheric chemistry cycles and hence atmospheric radiative forcing.

Health Effects: Standards and Guidelines

4. Carbon monoxide is a colourless and odourless gas which at very high concentrations can lead to severe poisoning. This results in loss of consciousness or, at sufficiently elevated concentrations, death. The effects of carbon monoxide arise from interference of the gas with the processes which transport oxygen around the body and by blocking essential biochemical reactions in cells. Exposure of humans to carbon monoxide in the air is manifest in raised levels of carboxyhaemoglobin in the blood, which replaces the normal oxygen carrier haemoglobin in red blood cells.

5. At lower levels of exposure this action of carbon monoxide can lead to effects on the heart and on the brain. People who suffer from coronary artery disease, and who are subject to bouts of angina are likely to be at risk if their oxygen transport is impaired. Equally, mental activity could be affected by reductions in oxygen supply resulting from exposure to carbon monoxide. Such changes could affect hand-eye co-ordination for example.

6. EPAQS has recommended a standard for carbon monoxide of 10 ppm as a running 8-hour average. This is essentially the same as the most stringent WHO Guideline for carbon monoxide. (In practice the EPAQS value is marginally more stringent than that of WHO since the former refers to running 8-hour averages whereas the WHO value is calculated over fixed 8-hour periods).

Current Air Quality

7. Since July 1972, measurements of hourly average carbon monoxide concentrations have been performed with continuous analysers at over 20 fixed-point locations in the UK. The Department of the Environment (DoE) funded the operation of 18 of these sites, while the Greater London Council (GLC) and later London Scientific Services (LSS) operated a further 4 sites in London.

8. Most of the measurements performed in the 1970s were made at the London Victoria urban background site and on the kerbside of five cities.

During the mid-1980s, the DoE monitoring network centred on the London sites. Since 1989, the size and geographic spread of the network has rapidly expanded in two stages. Monitoring was initially carried out at seven sites. Following the expansion of the urban monitoring network in 1991/92, the number of sites monitoring carbon monoxide has rapidly expanded to the position where, as of January 1997, there are currently 41 operational with a further 17 planned in the next year or so.

9. The early monitoring of carbon monoxide in the UK was carried out using a technique (non-dispersive infra-red absorption (NDIR)) which has subsequently been shown to be subject to interferences and appears to have produced data which were high by about 1-2 ppm on average. The only site with a sufficiently long run of data to provide trend information is the Central London site and the data are shown in Figure II.5.2. Although for the reasons already given the data from 1972 to 1979 are suspect, the more recent data from 1980 to 1991 show, overall, for both annual mean and 98th percentile of hourly means an upward trend of some 2% - 3% per year. Over the same period the UK-wide rate of increase in carbon monoxide emissions from 1980 to 1991 was about 6.5%.

Figure II.5.2:
Annual mean and 98th
percentile of carbon
monoxide in Central
London (1972-1995)

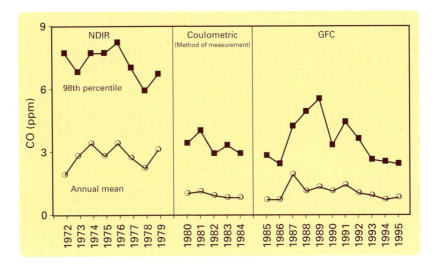

10. The current situation in terms of ambient carbon monoxide concentrations is shown in Table II.5.2. There is some indication that the 1995 values are lower than in previous years, which is consistent with overall estimates of emissions in Table II.5.1, resulting from the increasingly stringent vehicle emission limits imposed in recent years, although year-to-year meteorological variability could not be ruled out at this stage.

11. Although there is no very clear evidence of a downward trend in urban carbon monoxide concentrations, there is evidence that rural concentrations have begun to decrease. Measurements on the west coast of Ireland have shown clear decreases of about 28 ppb/year or 13%/year in carbon monoxide concentrations in polluted air masses arriving from the European continent over the last few years. This has been attributed to the lower emission motor vehicle technologies which have increasingly been used in Europe.

12. From knowledge of the sources of carbon monoxide and other pollutants, obtained from the development of emission inventories and modelling on local and national scales, carbon monoxide concentrations can be estimated for the whole of the UK based on national diffusion tube

Chapter II.5: Carbon Monoxide

Site	Year	Annual Average	Maximum 1-hour	Maximum 8-hour running average	No. of days the recommended EPAQS standard was exceeded
Stevenage	1990	0.7	6.4	3.5	0
	1991	0.7	7.4	4.9	0
	1992	0.6	5.0	3.0	0
	1993	0.5	6.8	3.0	0
	1994	–	–	–	–
London, Victoria	1990	1.3	11.7	10.2	1
	1991	1.4	13.9	11.2	3
	1992	1.0	9.3	6.6	0
	1993	0.9	10.6	7.5	0
	1994	0.7	9.6	8.9	0
	1995	0.8	6.6	5.6	0
London Cromwell Road (kerbside)	1990	2.9	18.4	15.5	5
	1991	3.3	18.7	13.9	5
	1992	2.8	11.0	8.7	0
	1993	2.2	12.6	9.1	0
	1994	1.9	10.7	10.1	1
	1995	1.8	9.7	6.7	0
West London	1990	1.5	13.2	10.2	3
	1991	1.7	18.0	15.8	4
	1992	1.1	8.7	6.3	0
	1993	0.9	12.4	8.7	0
	1994	0.8	10.8	8.8	0
	1995	0.7	14.1	11.1	2
London Bloomsbury	1992	0.8	5.8	4.5	0
	1993	0.6	5.3	3.5	0
	1994	0.6	8.3	6.5	0
	1995	0.6	6.3	4.3	0
Glasgow	1990	1.1	17.0	11.6	2
	1991	1.4	16.3	12.5	3
	1992	1.2	11.2	8.7	0
	1993	0.9	8.5	6.0	0
	1994	0.8	7.9	5.4	0
	1995	0.7	9.1	4.7	0
Manchester	1991	0.9	8.6	5.7	0
	1992	0.9	15.7	12.5	2
	1993	0.6	5.5	3.6	0
	1994	0.5	17.8	9.9	0
	1995	0.4	8.6	7.3	0
Sheffield	1991	0.9	8.3	6.3	0
	1992	0.7	9.3	7.4	0
	1993	0.4	6.7	4.5	0
	1994	0.4	5.9	4.3	0
	1995	0.7	6.6	4.4	0
Belfast Centre	1992	0.7	18.1	10.3	1
	1993	0.7	14.5	10.6	1
	1994	0.7	16.3	13.4	2
	1995	0.6	16.3	14.0	3

Table II.5.2:
continued

Site	Year	Annual Average	Maximum 1-hour	Maximum 8-hour running average	No. of days the recommended EPAQS standard was exceeded
Birmingham Centre	1992	0.6	14.2	10.8	1
	1993	0.6	4.5	3.8	0
	1994	0.6	14.1	9.7	0
	1995	0.6	9.2	6.9	0
Cardiff Centre	1992	0.7	9.6	4.9	0
	1993	0.6	9.7	5.9	0
	1994	0.8	7.2	5.7	0
	1995	0.5	7.2	3.6	0
Edinburgh Centre	1992	0.9	6.1	3.7	0
	1993	0.6	9.1	4.8	0
	1994	0.6	5.5	3.5	0
	1995	0.6	4.8	3.2	0
Newcastle Centre	1992	0.8	8.2	4.0	0
	1993	0.7	11.8	6.1	0
	1994	0.6	4.6	3.2	0
	1995	0.6	6.6	3.5	0
Leeds Centre	1993	0.8	10.6	8.9	0
	1994	0.7	12.8	9.5	0
	1995	0.7	7.2	5.5	0
Bristol Centre	1993	0.8	6.6	4.2	0
	1994	0.7	8.7	5.6	0
	1995	0.6	6.8	4.6	0
Liverpool Centre	1993	0.5	4.2	3.2	0
	1994	0.6	5.8	3.2	0
	1995	0.4	4.2	2.2	0
Birmingham East	1993	0.4	3.6	1.4	0
	1994	0.5	13.8	11.9	2
	1995	0.5	13.9	9.0	0
Leicester	1994	0.6	8.8	5.9	0
	1995	0.5	10.1	7.9	0
Southampton	1994	0.8	10.4	8.6	0
	1995	0.8	10.5	6.3	0
Hull	1994	0.6	10.1	5.2	0
	1995	0.6	5.8	3.9	0
Bexley	1994	0.5	8.4	4.5	0
	1995	0.4	6.2	4.7	0
Swansea	1995	0.5	9.1	3.7	0
Middlesbrough	1995	0.3	6.4	2.9	0

- Site not operational

Chapter II.5: Carbon Monoxide

**Figure II.5.3:
Estimated annual mean
carbon monoxide
concentrations 1994 (ppb)**

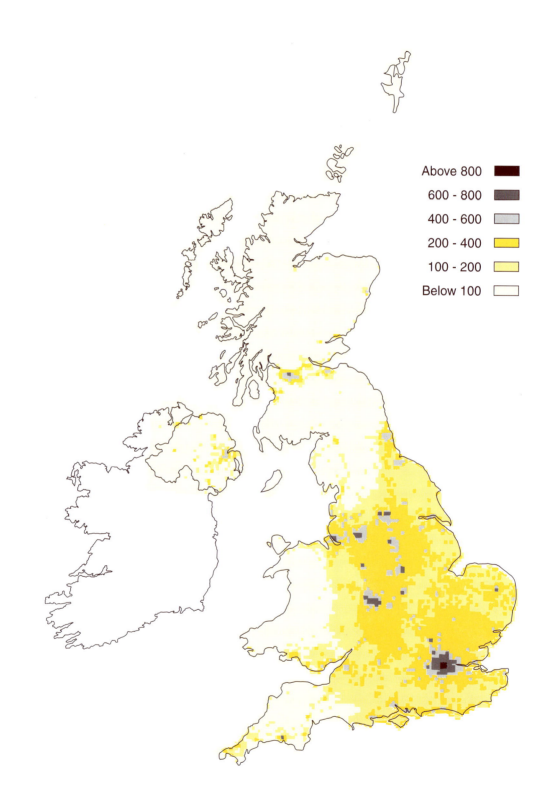

Above 800

600 - 800

400 - 600

200 - 400

100 - 200

Below 100

surveys of nitrogen dioxide. The map of current 1994 mean carbon monoxide concentrations is shown in Figure II.5.3. The density of traffic sources in the major urban areas is clearly apparent. However it should be noted that this map displays concentrations averaged over 5 by 5 kilometre OS grid squares which are more appropriate to urban background levels than roadside locations. While it is not appropriate or practicable to map such 'hot spots' at the national level, it should be recognised that within these grid squares higher levels of carbon monoxide will occur near busy roads.

13. Table II.5.2 also shows the extent of exceedence of the carbon monoxide standard. It can be seen that exceedences of the standard have occurred at most of the sites for which more than one year's data are available. The number of days exceeding the recommended standard ranges from 0-5 and the greatest margin of exceedence is 58%, although at most sites the margin of exceedence is less than 25%. The frequency and magnitude of the exceedences tend to be highest in the larger cities.

14. Exceedences of the level recommended by EPAQS tend to be greatest at the roadside, and this is consistent with the emission inventories showing motor vehicles as the dominant source. However, during severe pollution episodes such as the one which occurred in December 1991 in London, accumulation of pollutant across an urban area begins to dominate over local influences and concentrations may not be highest at the roadside. This was the case in 1991 when the kerbside Cromwell Road site showed a maximum 8-hour mean carbon monoxide of 13.9 ppm while the nearby urban background West London site showed a maximum of 15.8 ppm. In these conditions, the overall level of motor vehicle emissions in the surrounding area, rather than the proximity to busy roads is the major determinant of exposure.

The Strategy

15. The most important measures to reduce future carbon monoxide concentrations will inevitably be those which control emissions from petrol-engined vehicles. Measures already in place are the EC Directive 91/441/EEC, which has effectively required three-way catalysts to be fitted to new cars from January 1993, and further reductions from 1997 which set limits on carbon monoxide emissions which are 30% lower than 91/441/EEC.

16. Estimates of future emissions can be made on the basis of policy measures such as these, and incorporating the forecasts of national road traffic activity published by the Department of Transport. These suggest that national emissions of carbon monoxide will reduce by 32% in 2000, 48% in 2005 and by 54% in 2010 compared with 1995 levels.

17. If we assume that urban concentrations will reduce in the same proportions, then estimates of future carbon monoxide concentrations can be made and these are shown in Table II.5.3. This assumption may be pessimistic in that the extent of traffic growth in central urban areas (where carbon monoxide levels tend to be highest) may be smaller than that assumed in the national emission forecasts.

18. Nonetheless, the estimates in Table II.5.3 suggest that by 2000, the EPAQS standard will be met everywhere in the UK, with the possible exception, in some years, of the near vicinity of heavily trafficked roads. By 2010 there should be no exceedences of the EPAQS recommended standard. Beyond this date, there should also be no exceedences of the

Site	2000	2005	2010
Stevenage	2.0 – 3.3	1.6 – 2.5	1.4 – 2.2
London, Victoria	3.8 – 7.6	2.9 – 5.8	2.6 – 5.1
London, Cromwell Road	4.3 – 10.5	3.3 – 8.1	2.9 – 7.1
London, Earls Court	2.8 – 10.7	2.1 – 8.2	1.9 – 7.3
London, Bloomsbury	2.4 – 4.4	1.8 – 3.4	1.6 – 3.0
Glasgow	3.2 – 8.5	2.4 – 6.5	2.2 – 5.7
Manchester	2.4 – 8.5	1.9 – 6.5	1.6 – 5.7
Sheffield	2.9 – 5.0	2.2 – 3.8	2.0 – 3.4
Belfast Centre	7.0 – 9.5	5.3 – 7.3	4.7 – 6.4
Birmingham Centre	2.6 – 7.3	2.0 – 5.6	1.7 – 5.0
Cardiff Centre	2.4 – 4.0	1.9 – 3.1	1.7 – 2.7
Edinburgh Centre	2.2 – 3.3	1.7 – 2.5	1.5 – 2.2

Table II.5.3:
Projected maximum 8-hour carbon monoxide running mean at UK sites (ppm)

standard, provided traffic activity does not increase to the extent that the benefit from increasingly stringent emission limits are eroded.

Conclusions

19. The Government has decided to accept 10 ppm as a running 8-hour mean as the standard for carbon monoxide, and to adopt the objective of achieving the standard by 2005.

20. The Government has decided that the objective should apply both in the following non-occupational near-ground level outdoor locations: background locations; roadside locations; and other areas of elevated carbon monoxide concentrations where a person might reasonably be expected to be exposed (e.g. in the vicinity of housing, schools or hospitals etc) over a period of 8 hours. The latter areas are likely to represent an upper limit of exposures over 8 hour periods. Consideration of potential exceedences of the carbon monoxide objective at potentially high concentration locations should be carried out in conjunction with data from urban or other appropriate background locations. These conclusions will be reviewed in the light of further research on personal exposure, particularly insofar as indoor exposures are relevant, although it should be made clear here that the Strategy is not intended to protect smokers or those with combustion sources in their homes.

Chapter II. 6: Lead

Introduction and Sources

1. Lead is the most widely used non-ferrous metal and has a large number of industrial applications, both in its elemental form and in alloys and compounds. The single largest use globally is in the manufacture of batteries (60-70% of the world consumption of some 4 million tonnes of lead), but other uses are as a pigment in paints and glazes, in alloys, in radiation shielding, tank lining and piping. As the compound tetraethyl lead, it has been used as a petrol additive to enhance the octane rating. With the recognition of the adverse effects of lead on human health and the growing use of catalytic converters, which are poisoned by lead, this use is declining rapidly.

2. Most of the airborne emissions of lead in the UK arise from petrol engined motor vehicles. A summary of the UK inventory of emissions of lead is given in Table II.6.1.

3. Most of the lead in the air is in the form of fine particles with an aerodynamic diameter of less than 1 micron (1 micron is one millionth of a metre). In the immediate vicinity of smelters, the particle size distribution usually shows a predominance of larger particles. However, these particles settle out of the air at distances of a few hundred metres or 1-2 kilometres, so that further away from these sources the particle size distribution is similar to that at urban sites mainly influenced by traffic emissions of lead.

4. Direct human exposure to lead occurs through food, water, dust and soil, and air. Most people receive the largest portion of their daily lead intake via food, although other sources may be important in specific populations (e.g. water in areas with lead pipes and plumbosolvent water supply; air in populations living near point sources of lead; soil, dust and paint flakes in young children living in houses with leaded paint or contaminated soil). The percentage of lead absorbed from the gastrointestinal tract is about 10% in adults, and 40-50% in children. Absorption through the respiratory tract ranges from 20% to 60%. Children are also likely to be more susceptible to lead and may be at particular risk if they have a deficient intake of calcium, iron or vitamin D.

5. Lead exhibits toxic biochemical effects in humans which are manifest in the synthesis of haemoglobin, acute or chronic damage to the nervous system, effects on the kidneys, gastrointestinal tract, joints and reproductive system. These problems are well-described in workers exposed to high concentrations. Also some have been observed as a consequence of ingestion of lead, especially by young children. In conditions of low-level and long-term lead exposure such as are found in the general population, the most critical effects are those on haem biosynthesis, erythropoiesis, the nervous system, and blood pressure.

6. Most studies of the adverse effects of lead are based on blood lead levels. On absorption, lead is rapidly distributed through the body, where it accumulates. Typically about 2% of the body burden is in the blood where it is most biologically active. Its half life in blood is about three weeks. The rest is stored in bone (>90%), teeth, skin and muscle, where it is slowly released into the blood and potentially available for excretion via the kidneys.

7. In terms of exposure to lead, the consensus is that long-term exposure (over periods of the order of a year or more) is the relevant measure and this will be discussed further below in relation to air quality standards.

8. During the 1970s and early 1980s the lead content of petrol was gradually reduced, maintaining total emissions from vehicles broadly constant. At the end of 1985, the maximum permitted lead content of petrol was reduced significantly from 0.40 g/l to 0.15 g/l, and in 1987 unleaded petrol was introduced.

9. Emissions of lead from petrol vehicles are estimated annually and the results from 1980 to 1995 are shown in Figure II.6.1. The large increase in unleaded petrol consumption after 1988 has been responsible for more than halving emissions from motor vehicles since 1987.

Table II.6.1: Emissions of lead in the UK, 1995

Source	Emission (tonnes)	Percentage of total* in 1995
Road Transport: Petrol	1067	72
Road Transport: Diesel	1	0
Non-Ferrous Metals	140	9
Iron and Steel	46	4
Waste Related Sources	105	7
Industrial Processes	23	2
Industrial Combustion	44	3
Power Stations Combustion	28	2
Domestic Combustion	9	1
Other Combustion	5	0
Total	**1492**	**100**

* Figures rounded to nearest 1 %

Health Effects: Standards and Guidelines

10. The effects of lead on human health referred to above have generally been quantified by using the concentration of lead in blood, or blood-lead, as the indicator of exposure. Contributions to blood-lead levels arise primarily from air, water and food.

11. A brief qualitative description of the adverse effects of lead on human health was given above. The World Health Organisation has recently reviewed the literature on this subject and summaries of the lowest observable adverse effect levels (LOAELs) in adults and children are given in Tables II.6.2 and II.6.3.

12. Anaemia occurs only in cases of severe lead poisoning, but effects on red cell survival and haemoglobin production are found at lower levels. The lowest observed effect level at which impaired haem synthesis is found is at a blood lead concentration of about 50 µg/dl.

13. Acute neurological effects of delirium, confusion and convulsions are very rare and occur at blood lead levels above 100 µg/dl. More chronic effects include wrist drop and impaired mental function. Peripheral nerve dysfunction has been detected at levels above about 30 µg/dl. Subtle effects on neuropsychological function may be found in children at blood lead levels below 10 µg/dl. It is worth noting in this context that the Committee on Toxicity of Chemicals in Food, Consumer Products and the

Environment (CoT) has concluded that it is not possible to identify a threshold for effects of lead on health.

14. Workers are subject to biological monitoring at frequencies dependent on their blood lead levels. Standards for airborne lead in the workplace are 100 μg/m³ 8-hour time-weighted average for tetra-ethyl lead (the anti-knock additive used in leaded petrol) and 150 μg/m³ 8-hour time-weighted average for other forms of lead.

15. An ambient air quality standard exists within the EU via Directive 82/884/EEC and is 2 μg/m³ as an annual average. In their 1987 publication, the WHO set a guideline for lead at 0.5-1.0 μg/m³ also as an annual average, but have recently revised this guideline to 0.5 μg/m³ as an annual average. This was based on the target of ensuring that at least 98% of the exposed population should have blood lead levels below 10 μg/dl. On this basis the median blood lead level would not exceed 5.4 μg/dl. Using the relationship between blood lead and airborne lead, namely that 1 μg/m³ of airborne lead would contribute to 5 μg/dl blood lead (to allow for uptake by other routes), and that the maximum non-anthropogenic level of blood lead is 3μ g/dl, the WHO arrived at the value of 0.5 μg/m³ (or 500 ng/m³) as the guideline for an annual average for airborne lead.

Current Air Quality

16. Lead concentrations in air have decreased significantly in the past decade. Indeed, apart from the very large decrease in black smoke and sulphur dioxide concentrations following the Clean Air Act of 1956, it is arguable that the decline in airborne lead levels has been the next most dramatic change in pollutant levels in the UK in recent years.

17. The reason for this decrease is twofold. The major reduction in the maximum permissible lead content of leaded petrol from 0.4 μg/l to 0.15 μg/l in January 1986 almost halved urban air lead levels in the space of a few months. This reduction was reinforced and sustained by the introduction of unleaded petrol in 1987 and the continued increasing market share of this fuel ever since, to the point where, since 1993, all new petrol engined cars are catalyst equipped and therefore must run on unleaded petrol.

18. In the mid-1980s, annual average urban levels of airborne lead were broadly in the range 0.15-1.0 μg/m³, with one measurement in 1985 in

**Figure II.6.1:
Emissions of lead from
petrol vehicles**

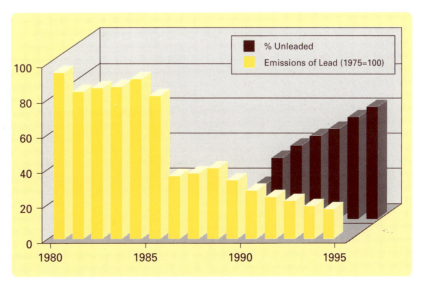

Cardiff of 1.28 µg/m³, and one in Manchester, also in 1985, of 2.04 µg/m³. Kerbside levels at the Cromwell Road site in West London were of the order of 1.4 µg/m³. Following the reductions in the lead content of petrol discussed above, urban levels have reduced to the extent that the maximum values are now of the order of 0.2-0.3 µg/m³ even at the Cromwell Road site. Rural values, as expected, are rather smaller and currently range from about 0.005-0.050 µg/m³. In industrial areas in the vicinity of processes which emit lead, such as secondary non-ferrous metal smelters, levels can be higher than in urban areas where motor vehicle emissions are the main source of lead. Levels at such sites in 1995 ranged from about 0.14-1.0 µg/m³.

19. The WHO guideline figure is therefore currently unlikely to be exceeded in urban areas. This is not the case for some industrial areas however, where levels, at the Walsall sites for example, exceed the WHO guideline, although they remain within the current EC Directive limit.

The Strategy

20. Unleaded petrol currently has a market share of some 70% of petrol sales. The increasing penetration of catalyst-equipped cars into the fleet will mean a progressive increase in this proportion, to the extent that emissions of lead from petrol vehicles should decrease by over 80% by 2005 compared with 1995 levels, and to a very small amount by 2015. Levels of airborne lead will therefore continue to decrease in urban areas where traffic is the major source, and, in the great majority of the UK, annual average levels should decrease to the extent where in most urban

Table II.6.2 : Summary of Lowest Observable Adverse Effect Levels for lead-induced health effects in adults

Lowest-observed-effect levels of blood lead (µg/l)	Haem synthesis and haematological effects	Effects on nervous system
1000–1200		Encephalopathic signs and symptoms
800	Frank anaemia	
500	Reduced haemoglobin production	Overt subencephalopathic neurological symptoms, cognition impairment
400	Increased urinary Ala and elevated coproporphyrin	
300		Peripheral nerve dysfunction (slowed nerve condition velocities)
200–300	Erythrocyte protoporphyrin elevation in males	
150–200	Erythrocyte protoporphyrin elevation in females	

Source: *Update and Revision of the Air Quality Guidelines for Europe*, WHO, 1994

areas levels are of the order of 0.1-0.2 µg/m³ even at busy roadsides. The revised WHO guideline of 0.5 µg/m³ should therefore be achievable by 2005 in urban areas of the UK.

21. In the vicinity of individual industrial plants which may be significant emitters of airborne lead this may not be the case. Such plants will be subject generally to BATNEEC and the provisions of the Environmental Protection Act 1990 and emissions should be reduced to the lowest level possible in accordance with BATNEEC.

Conclusion 22. The Government recognises that EPAQS has not yet made a recommendation for a standard for lead. In the meantime, the Government will adopt the revised WHO guideline figure of 0.5 µg/m³ as an annual mean, as the standard, with the objective of achieving it by 2005. Noting the advice of the Committee of Toxicology, the Government further proposes to reduce levels of airborne lead below this target level wherever practicable.

23. The Government has decided that the objective for lead should apply in the following non-occupational near-ground level outdoor locations: background locations; roadside locations; and other areas of elevated lead concentrations where a person might reasonably be expected to be exposed (e.g. in the vicinity of housing, schools or hospitals etc), over the averaging time of the objective. Consideration of potential exceedences of the lead objective at potentially high concentration locations should be carried out in conjunction with data from urban or other appropriate background locations. It should be noted that exposure to lead is from a variety of sources, and it is UK Government policy to reduce exposure to lead from all sources including food, drinking water, paint, air, soil and dust, and thus restrict as far as possible the environmental accumulation of lead.

Table II.6.3: Summary of Lowest Observable Adverse Effects Levels for lead induced health effects in children

Lowest-observed-effect levels of blood lead (µg/l)	Haem synthesis and haematological and other effects	Effects on nervous system
800–1000		Encephalopathic signs and symptoms
700	Frank anaemia	
400	Increased urinary ALA and elevated coproporphyrin	
250–300	Reduced haemoglobin synthesis	
150–200	Erythrocyte protoporphyrin elevation	
100–150	Vitamin D3 reduction	Cognitive impairment
100	ALAD – inhibition	Hearing impairment

Source: *Update and Revision of the Air Quality Guidelines for Europe*, WHO, 1994.

Table II.6.4: Annual mean airborne lead concentrations at UK sites 1980-1995 (ng/m³)

	1980	1981	1982	1983	1984	1985	1986	1987	1988	1989	1990	1991	1992	1993	1994	1995
KERBSIDE																
Cromwell Road (W. London)	-	-	-	1,370	1,410	1,450	660	-	-	-	380	360	348	255	244	199
URBAN																
Central London	640	580	630	470	520	480	270	280	300	220	-	120	99	78	85	60
Brent (London)	770	710	890	990	-	640	300	290	320	-	220	200	174	147	144	80
Leeds	650	370	450	440	260	310	180	190	140	-	120	-	-	106	80	76
Motherwell	260	230	300	240	180	260	190	180	-	-	200	160	50	86	23	50
Glasgow	460	330	240	420	190	270	120	180	130	140	95	92	93	90	39	51
Cardiff	-	-	-	-	-	1,280	630	630	620	570	460	440	384	311	233	165
Manchester	-	-	-	-	-	2,040	810	810	760	640	510	460	339	305	123	133
Newcastle	-	-	-	-	-	180	130	150	110	110	70	70	67	70	27	25
North Tyneside	-	-	-	-	-	290	150	190	140	120	81	100	90	90	26	-
RURAL																
Cottered	-	-	-	-	130	130	77	98	76	75	41	45	44	36	19	20
North Petherton	-	-	-	-	-	-	70	65	69	81	53	62	-	-	-	-
Eskdalemuir	-	-	-	-	-	29	8	13	14	13	10	13	10	7	6	5
Chilton	110	56	66	65	86	90	33	60	52	48	38	38	27	22	25	25
Trebanos	82	80	78	92	98	81	39	48	55	64	43	63	60	37	32	39
Styrrup	178	135	172	115	170	130	66	70	94	86	57	65	55	35	47	44
Windermere	47	39	47	45	48	35	24	20	23	21	15	20	8	8	12	13
INDUSTRIAL																
Walsall Industry 1	-	-	-	-	-	-	930	1,370	1,160	880	470	570	540	570	500	697
Walsall Metals Industry 2	-	-	-	-	-	-	2,260	2,950	3,580	2,430	1,300	1,390	1,440	1,220	1,400	1,020
Walsall Metals Industry 3	-	-	-	-	-	-	760	730	-	640	-	340	410	-	-	-
Walsall Metals Industry 5	-	-	-	-	-	-	-	1,110	1,590	1380	680	620	680	470	470	660
Brookside 1	-	-	-	-	-	-	310	330	260	260	150	150	220	160	150	177
Brookside 2	-	-	-	-	-	-	990	890	1,310	1,100	1,140	710	560	470	440	464
Elswick 1	-	-	-	-	-	-	-	1,650	1,350	530	370	660	340	190	340	481
Elswick 2	-	-	-	-	-	-	-	610	510	600	330	250	230	150	150	141
Elswick 6	-	-	-	-	-	-	-	-	-	780	450	460	300	190	220	192

Chapter II.7: Nitrogen Dioxide

Introduction and Sources

1. Nitrogen dioxide (NO_2) and nitric oxide (NO) are both oxides of nitrogen and together they are referred to as NO_x. All combustion processes in air produce NO_x, though the colourless gas NO usually predominates. The conversion of NO to the red-brown gas NO_2 takes place in the atmosphere via reaction with chemically active species such as ozone. Generally in remote, unpolluted areas nitric oxide concentrations are only a small fraction of those of nitrogen dioxide. However, in polluted areas where the oxidizing capacity of the atmosphere may be limited, nitric oxide concentrations often exceed those of nitrogen dioxide. It is nitrogen dioxide which is associated with adverse effects upon human health, particularly with regard to exacerbation of symptoms associated with respiratory illness. Nitrogen oxides are also indirect greenhouse gases which influence atmospheric chemistry cycles and hence atmospheric radiative forcing.

2. Current estimates show that, on a national scale, road transport accounts for 46% of the total UK emissions of nitrogen oxides. The Electricity Supply Industry accounts for approximately 22%, the industrial and commercial sectors for 12% and domestic sources for 3%. The remainder was emitted from a variety of other sources including shipping, railways and civil aviation (see Table II.7.1).

Table II.7.1: Emissions of NO_x in the UK, 1995

Source	Emissions kilotonnes*	Percentage of total in 1995**
Power stations	498	22
Domestic	66	3
Commercial/public service	35	2
Refineries	47	2
Iron and steel	48	2
Other industrial combustion	145	6
Non-combustion processes	2	<1
Extraction and distribution of fossil fuels	112	5
Road transport	1,062	46
Railways	21	1
Civil aircraft	16	1
Shipping	114	5
Military	41	2
Off-road	81	4
Waste treatment and disposal	5	<1
Agriculture	2	<1
Total	**2,293**	**100**

*   Figures rounded to nearest kilotonne
** Figures rounded to nearest 1%

Chapter II.7: Nitrogen Dioxide

Health Effects:
Standards and
Guidelines

3. Exposure to nitrogen dioxide can bring about reversible effects on lung function and airway responsiveness. It may also increase reactivity to natural allergens. Repetitive exposure in animals can produce changes in lung structure, lung metabolism and lung defences against bacterial infection. Animal toxicological studies suggest that peak concentrations contribute more to the toxicity than does the duration of the exposure although the latter is still important. Exposure to nitrogen dioxide may put children at increased risk of respiratory infection and may lead to poorer lung function in later life. There is some evidence to suggest that women exposed to raised concentrations of nitrogen dioxide as a result of using gas cookers and open gas fires may have impaired lung function.

4. Present EC legislation has set a limit value of 104.6 ppb (200 $\mu g/m^3$) expressed as a 98th percentile of hourly means in a calendar year. In addition, the Directive also provides for guide values of 70.6 ppb (135 $\mu g/m^3$) and 26.2 ppb (50 $\mu g/m^3$) expressed as 98th and 50th percentiles of the hourly mean values respectively.

5. A recommendation from the Expert Panel on Air Quality Standards (EPAQS) for nitrogen dioxide has recently been published[1]. A value of 150 ppb as an hourly mean has been recommended. In addition, EPAQS concluded that a longer-term standard is also desirable in order to protect against possible cumulative effects on the health of the population. However, they felt that there was insufficient evidence at the present time to set an appropriate figure but recommended a strategy to reduce annual average concentrations. The WHO guidelines have recently undergone revision such that there are now only two values. The first is a 1-hour guideline value of 200 $\mu g/m^3$ (104.6 ppb) and an annual average guideline of 40 $\mu g/m^3$ (20.9 ppb). The EPAQS recommended hourly standard is somewhat higher than the WHO hourly guideline. This difference depends on the intepretation of research findings published in 1976 which reported effects in people with asthma at concentrations as low as 100ppb. However, these findings have never been repeated in any other study and the Expert Panel considered that there was sufficient doubt about the validity of the nitrogen dioxide measurements for this isolated study to be disregarded.

Current Air Quality

6. The Department of the Environment operates, as of January 1997, a network of 56 automatic monitoring stations at various locations throughout the United Kingdom. These sites are situated in urban background, urban kerbside or rural locations and data is captured telemetrically on an hourly basis. Further expansion of the network is foreseen with an expectation that some 85 sites will be in operation by the end of 1997.

7. Table II.7.2 shows the summary information from the network for 1995. Of the 23 urban sites measuring nitrogen dioxide, 'poor air quality' (currently defined by the DoE as hourly concentrations greater than 100 ppb) was measured at 15 locations on up to 33 days of the year. The highest hourly concentrations measured at urban background locations were in London at Bridge Place (217 ppb), Belfast (199 ppb), Manchester (181 ppb) and Birmingham (177 ppb). These levels are likely to be

[1] Expert Panel on Air Quality Standards, *Nitrogen Dioxide*, 1996, The Stationery Office, London.

representative of outdoor exposure in these central urban areas. It is also interesting to note that the highest frequencies of poor air quality were recorded at the Central London kerbside site at Cromwell Road (33 days) and the background locations in Bridge Place (14 days), London

Table II.7.2 Nitrogen dioxide network summary (Annual statistics, calendar year 1995)

Site	Annual mean (ppb)	50th % (ppb)	98th % (ppb)	Max. hour (ppb)	No. of hours >= 100 ppb	On No. of days	
Cromwell Road (K)	47	44	101	170	178	33	
West London (U)	28	27	64	131	12	4	
Glasgow (U)	26	26	51	81	0	0	
Manchester (U)	23	21	60	181	16	7	
Walsall (U)	24	23	53	119	6	4	
Billingham (U)	18	16	48	90	0	0	
Sheffield Tinsley (U)	26	25	56	119	6	3	
Bridge Place (U)	34	31	79	217	65	14	
Lullington Heath (Rl)	8	6	25	43	0	0	
Strath Vaich (Rm)	1	0	3	11	0	0	
Ladybower (Rl)	8	6	27	40	0	0	
Harwell (Rl)	14	14	31	39	0	0	
London Bloomsbury (U)	35	34	71	176	36	8	
Edinburgh Centre (U)	26	26	54	120	4	3	
Cardiff Centre (U)	22	21	47	86	0	0	
Belfast Centre (U)	21	19	51	199	15	3	
Birmingham Centre (U)	24	23	57	177	24	6	
Newcastle Centre (U)	21	20	45	91	0	0	
Leeds Centre (U)	26	26	53	109	1	1	
Bristol Centre (U)	25	23	62	118	4	3	
Liverpool Centre (U)	26	25	58	90	0	0	
Birmingham East (U)	22	20	54	170	27	5	
Hull Centre (U)	24	23	50	93	0	0	
Leicester Centre (U)	23	23	48	106	3	1	
Southampton Centre (U)	24	23	51	95	0	0	
Bexley (U)	22	21	52	132	1	1	
Swansea (U)	22	21	50	94	0	0	
Middlesbrough (U)	17	15	42	97	0	0	

Key: K = Kerbside S = Suburban Rm = Remote U = Urban Rl = Rural

Bloomsbury (8 days) and Birmingham (6 days). Annual mean concentrations of nitrogen dioxide in urban locations for 1995 range from 17 ppb in Middlesbrough to 47 ppb measured at the kerbside site at Cromwell Road.

8. Statistically significant reductions in both the annual mean and the 98th percentile of the measured hourly values of nitrogen dioxide have been observed for the sites at Glasgow, Walsall, West London and Billingham over the past nine years. The data for the kerbside site at Cromwell road shows a decline of 5 ppb/year in the 98th percentile value over the last 14 years.

Table II.7.3: Summary of statistically significant trends for Nitrogen Dioxide and NO$_x$ (long-running trends with ten years of measurements are highlighted)

Pollutant	Site	Annual Parameter	Start Year	End Year	Slope	Slope error	Units
NO$_2$							
	Cromwell Rd	98th percentile	82	95	5.0	2.0	ppb/yr
	West London	98th percentile	87	95	−4.0	1.0	ppb/yr
	West London	Mean	87	95	−1.2	0.3	ppb/yr
	Glasgow	Mean	87	95	−0.6	0.2	ppb/yr
	Glasgow	98th percentile	87	95	−1.8	0.5	ppb/yr
	Billingham	Mean	87	95	−0.8	0.2	ppb/yr
	Billingham	98th percentile	87	95	−2.5	0.5	ppb/yr
	Bridge Place	Mean	91	95	−2.3	0.7	ppb/yr
	Walsall	Mean	87	95	−0.6	0.2	ppb/yr
	Walsall	98th percentile	87	95	−1.1	0.4	ppb/yr
NO$_x$	Central London	98th percentile	77	95	+5.0	2.0	ppb/yr
	Cromwell Rd	98th percentile	82	95	−19.0	7.0	ppb/yr
	Cromwell Rd	98th percentile	91	95	−60.0	20.0	ppb/yr
	Cromwell Rd	Mean	91	95	−17.0	4.0	ppb/yr
	West London	Mean	87	95	−5.0	2.0	ppb/yr
	West London	98th percentile	87	95	−30.0	10.0	ppb/yr
	Glasgow	Mean	87	95	−1.8	0.8	ppb/yr
	Glasgow	98th percentile	87	95	−14.0	6.0	ppb/yr
	Walsall	Mean	87	95	−2.6	0.7	ppb/yr
	Walsall	98th percentile	87	95	16.0	6.0	ppb/yr
	Billingham	98th percentile	87	95	−8.0	3.0	ppb/yr
	Billingham	Mean	87	95	−1.4	0.4	ppb/yr
	Bridge Place	Mean	91	95	−8.0	2.0	ppb/yr
	Bridge Place	98th percentile	87	95	−40.0	10.0	ppb/yr

However, four of the earlier years at this site had limited data capture (i.e. less than 50%) and thus the trend should be treated with caution. When total NO_x concentrations are analysed then statistically significant reductions have been observed in the 98th percentile values at the central London site (for the years 1987 to 1995), and the sites at Cromwell Road, West London, Glasgow, Bridge Place, Walsall and Billingham. Some of these sites are particularly affected by local traffic and it may be that changes in local emissions due to pedestrianisation schemes (at Glasgow) and other traffic management measures (at Cromwell Road) may be responsible for the observed decreases. Moreover, the trend data for the West London site may be skewed by the abnormally high values recorded in 1989, when a technical breach of the EC Limit Value occurred due to interference by a faulty exhaust ventilation system very close to the monitor. A significant increase in the 98th percentile hourly NO_x value has also been observed at the Central London site over a period of 19 years. A summary of the long term trends in the measured NO_x concentrations is given in Table II.7.3. Overall therefore, it is difficult to detect any clear trends, either up or down, in nitrogen dioxide concentrations in central urban areas across the UK.

9. Data from a much larger number of sites and range of locations than the data set discussed above are available from nitrogen dioxide measurements with passive diffusion tube samplers which were undertaken in 1986 and 1991 at over 360 sites within the UK. The samplers were located at existing smoke and sulphur dioxide monitoring stations and covered a wide range of town and city sizes and urban locations. In order to improve the spatial coverage of the measurements and to monitor the effects of Government policies on nitrogen dioxide concentrations (eg. introduction of three-way catalytic converters on new petrol cars) a new enlarged long-term survey was instigated in 1993. In 1995, 1220 sites were in operation and data capture was sufficient so as to allow annual means to be derived at 1116 sites.

10. Each participating local authority (approximately 300) operates 4 newly selected diffusion tube monitoring sites: one kerbside, one intermediate and two in background locations. Monthly average nitrogen dioxide concentrations for all sites are sent by each authority to the National Environmental Technology Centre (NETCEN), who publish annual reports of the acquired data. The most recent data is for 1995[2]. The highest annual average kerbside measurements were at Hackney (52 ppb, 99 $\mu g/m^3$) and Manchester (45 ppb, 86 $\mu g/m^3$) which are at least twice the lower limit of the recommended revised WHO guideline of 20.9 ppb (40 $\mu g/m^3$).

11. Average urban concentrations of nitrogen dioxide measured from the diffusion tube survey (during the period July to December of each year) showed an increase of approximately 34% over the period 1986 to 1991. The observed increases occurred throughout the country and were not confined to any particular area or region although there was a tendency for the ratio to be larger in the north and west of the country. During this period urban emissions of nitrogen oxides from road transport increased by about 30% which is consistent with the measured increase in nitrogen dioxide concentrations. It is still too early to undertake any analysis of trends in the current phase of the diffusion tube survey.

[2]*UK Nitrogen Dioxide Survey 1995*, National Environmental Technology Centre. Culham, Oxfordshire, UK. 1997.

[3]The Quality of Urban Air Review Group. *Urban Air Quality in the United Kingdom*. 1993.

Table II.7.4: Summary of nitrogen dioxide exceedances for calendar year 1995

Site	DOE 'Poor' (100–299ppb) (No. hours)[a]	EC Directive Criteria Hourly means)			EPAQS/WHO Criteria	
		98%ile 104.6ppb	98%ile 70.6ppb	50%ile 26.2ppb	Hourly max. >150ppb (number)[a]	Annual >21ppb (ppb)
		(value in ppb)				
Cromwell Road (K)	178(33)		101	44	9(5)	47
West London (U)	12(4)			27		28
Glasgow (U)						26
Manchester (U)	16(7)				5(1)	23
Walsall (U)	6(4)					24
Billingham (U)						
Sheffield Tinsley (U)	6(3)					26
Bridge Place (U)	65(14)		79	31	5(3)	34
Lullington Heath (Rl)						
Strath Vaich (Rm)						
Ladybower (Rl)						
Harwell (Rl)						
London Bloomsbury (U)	36(8)		71	34	2(1)	35
Edinburgh Centre (U)	4(3)					26
Cardiff Centre (U)						22
Belfast Centre (U)	15(3)				6(3)	21
Birmingham Centre (U)	24(6)				4(2)	24
Newcastle Centre (U)						21
Leeds Centre (U)	1(1)					26
Bristol Centre (U)	4(3)					25
Liverpool Centre (U)						26
Birmingham East (U)	27(5)				8(2)	22
Middlesbrough (U)						
Hull Centre (U)						24
Leicester Centre (U)	3(1)					23
Southampton Centre (U)						24
Bexley (U)	1(1)					22
Swansea (U)						22

Key: K = Kerbside S = Suburban Rm = Remote U = Urban Rl = Rural

a – numbers in parenthesis refer to the number of days on which exceedences occurred

12. Taking the automatic monitoring results together with those from the diffusion tube surveys, the Quality of Urban Air Review Group (QUARG) concluded in its first report[3] that "there are few data to indicate long-term trends across the country, but those that are available suggest a general increase in nitrogen dioxide concentration over the last decade. However, recent data for a limited number of traffic-saturated inner-city sites have shown little change in concentrations."

13. There were no exceedances of the EC limit value in 1995 although the 98th percentile guide value was exceeded at Cromwell Road, London Bloomsbury and Bridge Place. The 50th percentile guide value was exceeded at 4 locations where there was a significant data capture, namely, Cromwell Road, West London, London Bloomsbury and Bridge Place. In Table II.7.4 a summary of all the exceedances is recorded for the EC Directive criteria and also for the revised World Health Organization (WHO) air quality guideline based on the annual mean and the EPAQS recommended standard based on hourly means.

14. The EPAQS recommended standard of 150 ppb as an hourly mean, was exceeded at 7 urban locations, including the kerbside site at Cromwell Road, on up to 5 days in the year. With the exception of Billingham all of the urban monitoring sites would have exceeded the revised WHO annual average guideline in 1995.

15. Results from the 1995 UK Diffusion Tube Survey show that there were 326 locations across the UK which exceeded WHO annual average guideline. The highest values were recorded at Hackney (52 ppb, 99 $\mu g/m^3$) and Manchester (45 ppb, 86 $\mu g/m^3$). Of the background and intermediate sites, 126 exceeded the revised WHO annual guideline where the maximum recorded value was 39 ppb (74 $\mu g/m^3$) at Hackney in London.

The Strategy 16. The London Research Centre's emissions inventory for Greater London[4] based on energy use shows that road transport accounted for 75.9% of all NO_x emissions in London in 1990. The remainder was derived from power generation (1.1%), domestic sources (6.4%), small industrial/commercial sources (12.9%), railways (0.7%), aviation (2.9%) and transport by water (0.1%). A recent inventory derived for the West Midlands, again compiled by the London Research Centre attributes some 85% of the emissions of nitrogen oxides to road transport. Further inventories for other urban areas are at present being compiled. They are also likely to show a very significant contribution from road transport to total NO_x emissions. The 1995 emissions of nitrogen oxides for the whole of the UK are shown in Table II.7.1.

17. It must be remembered that nitrogen oxides emitted from large stationary combustion sources are generally emitted from greater heights compared to the NO_x emitted from road traffic. This means that for a unit mass of emission, traffic sources contribute significantly more to ground level concentrations than do stationary combustion sources. Gaussian plume modelling studies have documented the validity of this difference for London[5]. It is expected that in central London stationary sources contribute no more than 20% to average ground level NO_x concentrations.

[4]*London Energy Study*, London Research Centre, 1993.

[5]P.K. Munday, R.J. Timmis, C.A. Walker, *A Dispersion Modelling of Present Air Quality and Future Nitrogen Oxides Concentrations in Greater London, LR 731 (AP)M*, Warren Spring Laboratory, 1989.

In the more extreme high-pollution episodes when emissions are contained within a low-level inversion, which may be as little as tens of metres deep, this contribution may be even smaller.

18. The more disparate nature of these stationary sources makes them more difficult to control relative to those emissions from road transport. The Science Policy Research Unit (SPRU) boiler study represents the best available data source on UK boiler stocks below 50 MW. The study shows that there are over half a million boilers below 1 MW rated power (one megawatt thermal capacity) which account for over 73 GW of capacity. In the range 1-5MW there are approximately 21,000 boilers accounting for 49 GW of capacity and 4,900 boilers in the range 5-50 MW accounting for 38 GW of capacity. In general terms the smaller boilers are more likely to be using cleaner fuels such as gas or gas oil. For example, of the half million boilers below 1 MW, 373,000 burn natural gas. Throughout this sector the rate of boiler replacement is very low (perhaps 2% per annum) and shut down of the oldest boilers is more likely to arise through company closure. It is not uncommon to find boilers which are 20-30 years old still using their original burners.

19. At present under Part B of the Environmental Protection Act 1990, local authorities only have control over boilers and furnaces in the range 20-50 MW (not aggregated). Local authorities also have control over waste incineration plant below 3 MW rated power. The SPRU study indicates that such boilers are usually aggregated together via a common stack and are thus controlled under Part A of the Environmental Protection Act 1990. It

Table II.7.5: Projections of future urban road transport NO$_x$ emissions (kilotonnes)

	1995	2000	2005	2010	2015
Base Casea	409.3	303.7	252.5	238.2	242.6
Stage 2000	409.3	303.7	223.5	189.1	181.5
Stage 2005	409.3	303.7	216.2	154.9	130.0
Objective (100 %ile EPAQS)		212.8	212.8	212.8	212.8
Objective (annual)		155.5	155.5	155.5	155.5

a – Base Case includes measures up to and including 'Euro II' see para 20 of this chapter

Figure II.7.1: Future UK urban road transport emissions of nitrogen oxides (current policies; diesel car sales @ 20% of all new car sales)

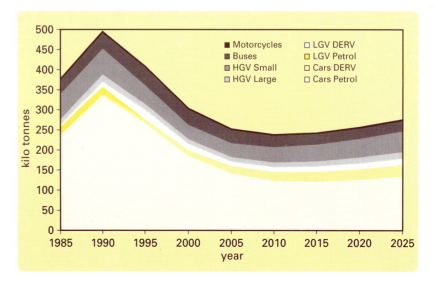

is unlikely that such boilers and incineration plant will be located in central urban areas. It therefore seems apparent that the major source of non-transport urban emitted NO_x is from space heating by small boilers which are presently uncontrolled and which are replaced only very slowly. As a consequence urban NO_x emissions from this sector are not expected to reduce significantly in the future.

20. This is not the case for transport emissions however, and Figure II.7.1 and Table II.7.5 shows how urban road traffic emissions of nitrogen oxides are expected to change. In making these forecasts the National Road Traffic Forecasts (NRTF) published by the Department of Transport have been used (in the projections presented here, the mean of the High and Low NRTF traffic activity projections has been used) together with the assumption that the following policies are implemented:

■ 'EURO I' vehicle emissions Directives which set limits for the emissions of nitrogen oxides, hydrocarbons, carbon monoxide and particulates (diesel vehicles only). In practice EURO I required the fitting of catalytic converters to all new petrol cars;

■ Directive 93/441/EEC, passenger cars (from 31/12/1992);

■ Directive 93/59/EC, light vans (from 1/10/1994);

■ Directive 91/542/EEC, heavy duty vehicles (from 1/10/1993);

■ 'EURO II' vehicle emissions Directives which tighten further the limits in EURO I;

■ Directive 94/12/EC, passenger cars (from 1/1/1997);

■ Directive 96/69/EC, light vans (from 1/10/97);

■ Directive 91/542/EEC, heavy duty vehicles (from 1/10/1996).

The target emissions are based upon the assumption that the required emissions reductions come only from the road transport sector.

Table II.7.6: Urban box modelling of the December 1991 pollution episode using historical and future road transport emissions scenarios

Year	Urban UK Road Transport NO_x Emissions (kilotonnes per annum)	Peak Hourly Mean NO_2 Concentration (ppb)
1970	260.2	186
1975	288.2	212
1980	337.2	263
1985	377.3	308
1990	495.1	453
1995	407.2	343
2000	289.6	213
2005	218.1	149
2010	202.7	138
2015	210.5	144
2020	226.0	156
2025	242.8	171

Chapter II.7: Nitrogen Dioxide

**Figure II.7.2:
The observed
relationship between
NO$_2$ and NO$_x$**

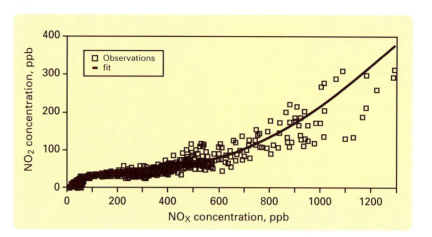

**Figure II.7.2:
The observed
relationship between
NO$_2$ and NO$_x$**

21. Diesel car sales are assumed to remain at 20% of total car sales from 1994 onwards (present day levels are at ca. 22%).

22. At this stage no account has been taken of the proposals on new vehicle emission standards proposed recently by the European Commission.

23. There are considerable difficulties in estimating future emissions as discussed in the chapter on particles. In the case of NO$_x$, there is some information on the degradation in emissions performance from work in the Netherlands and Germany, and this has been used here. The assumption is that the emissions performance of three-way catalyst cars degrades linearly with time until after 80,000 km the emissions are a factor of 1.6 times the regulated value. For post-2000 vehicles a similar approach is adopted except that the factor is 1.1 to take account of assumed improvements in inspection and maintenance. The extent to which this is likely to represent adequately the real in-service performance, either in the countries where the data originated or particularly in the UK, is unknown. Moreover, the projections use the mean of the high and low NRTF forecasts, and assume that the penetration of diesel cars into new car sales does not increase beyond the current value of about 20%.

**Short Period Air
Quality Targets**

24. These projections can be used to estimate the likelihood of attaining air quality objectives. The first part of this analysis looks at future emissions and air quality in relation to full compliance with the EPAQS recommended standard of 150 ppb (286.5 µg/m^3) as an hourly mean. There are sound reasons for assessing policies against the highest measured nitrogen dioxide concentration, since the highest value in the past few years occurred in the pollution episode of December 1991 in London. A study of that period concluded that it was associated with an increase in mortality and morbidity, and that air pollution was a plausible explanation for this increase. The study was not able to determine the relative contributions of black smoke and nitrogen dioxide, which were the main pollutants at the time.

25. The peak hourly concentrations observed during the 1991 episode were unprecedented. The episode took place when pressure was high for several days over western Europe, wind speeds were low and widespread freezing fog was reported. In stable conditions such as these vertical motion of the air is suppressed and cold dense air is liable to rest at the surface. The peak hourly concentration of nitrogen dioxide measured during this episode was 423 ppb (808 µg/m^3)

TABLE II.7.7 Maximum hourly mean concentrations of nitrogen dioxide (ppb) for the calendar years 1990-1995

Site	Year					
	1990	1991	1992	1993	1994	1995
Cromwell Road (K)	245	382	134	186	235	170
West London (U)	151	388	142	155	198	131
Glasgow (U)	114	213	94	109	132	81
Manchester (U)	191	127	369	164	352	181
Walsall (U)	153	137	269	118	257	119
Billingham (U)	125	130	143	108	112	90
Sheffield (U)	89	189	179	214	148	119
Bridge Place (U)	150	423	175	140	268	217
Lullington Heath (Rl)	36	54	90	53	64	43
Strath Vaich (Rm)	15	25	13	15	22	11
Ladybower (Rl)	48	60	57	54	62	40
London Bloomsbury (U)	-	-	141	120	207	176
Edinburgh Centre (U)	-	-	71	114	91	120
Cardiff Centre (U)	-	-	79	77	264	86
Belfast Centre (U)	-	-	130	186	100	199
Birmingham Centre (U)	-	-	127	262	136	177
Newcastle Centre (U)	-	-	285	271	204	91
Leeds Centre (U)	-	-	-	83	131	109
Bristol Centre (U)	-	-	-	86	91	118
Liverpool Centre (U)	-	-	-	108	101	90
Birmingham East (U)	-	-	-	43	189	170
Hull Centre (U)	-	-	-	-	111	93
Leicester Centre (U)	-	-	-	-	97	106
Southampton Centre (U)	-	-	-	-	81	95
Bexley (U)	-	-	-	-	106	132
Swansea (U)	-	-	-	-	82	94
Harwell (Rl)	-	-	-	-	-	39
Middlesbrough (U)	-	-	-	-	-	97

Key: K = Kerbside S = Suburban Rm = Remote U = Urban Rl = Rural

26. The relationship between measured NO_x and measured nitrogen dioxide concentrations is highly non-linear which makes analysis of the relationship between required reductions in nitrogen dioxide concentrations and those of NO_x difficult (see Figure II.7.2). However, using the empirical relationship between measured nitrogen dioxide

Chapter II.7: Nitrogen Dioxide

TABLE II.7.8: Annual
average concentrations of
nitrogen dioxide (ppb) for
the calendar years
1990-1995

Site	Year					
	1990	1991	1992	1993	1994	1995
Cromwell Road [K]	42	43	41	40	44	47
West London [U]	35	35	32	31	31	28
Glasgow [U]	26	26	25	27	26	26
Manchester [U]	28	27	31	26	26	23
Walsall [U]	25	26	27	24	25	24
Billingham [U]	21	19	18	16	17	18
Sheffield [U]	24	28	31	28	28	26
Bridge Place [U]	37	43	39	34	34	34
Lullington Heath [Rl]	6	8	7	9	9	8
Strath Vaich [Rm]	1	2	1	1	1	1
Ladybower [Rl]	13	11	9	10	9	8
London Bloomsbury [U]	–	–	39	34	34	35
Edinburgh Centre [U]	–	–	28	27	27	26
Cardiff Centre [U]	–	–	25	23	22	22
Belfast Centre [U]	–	–	23	22	21	21
Birmingham Centre [U]	–	–	24	25	24	24
Newcastle Centre [U]	–	–	27	28	25	21
Leeds Centre [U]	–	–	–	26	28	26
Bristol Centre [U]	–	–	–	26	24	25
Liverpool Centre [U]	–	–	–	27	26	26
Birmingham East [U]	–	–	–	17	20	22
Hull Centre [U]	–	–	–	–	24	24
Leicester Centre [U]	–	–	–	–	23	23
Southampton Centre [U]	–	–	–	–	23	24
Bexley [U]	–	–	–	–	22	22
Swansea [U]	–	–	–	–	–	22
Harwell [Rl]	–	–	–	–	–	14
Middlesbrough [U]	–	–	–	–	–	17

Key: K = Kerbside   S = Suburban   Rm = Remote   U = Urban   Rl = Rural

concentrations and those of NO_x, estimates of the required emission reductions in urban NO_x can be made, assuming that the contribution from road traffic remains at 80% and that contribution from other sources remains constant. For a nitrogen dioxide concentration of 423 ppb, this approach suggests that a reduction in urban road transport NO_x emissions of about 48% relative to 1995 would be required.

27. The maximum hourly nitrogen concentration measured in 1994 was 352 ppb (673 µg/m^3) recorded in Manchester. Thus the nitrogen dioxide concentration would have to be reduced by approximately 57% so as not to exceed the EPAQS hourly standard of 150 ppb. Figure II.7.2 can be used to translate this desired nitrogen dioxide reduction into a NO$_x$ emissions reduction. Using the same argument as above, it transpires that a reduction of about 32% from present day levels would have been required in urban emitted NO$_x$ in order to have met this guideline in Manchester. This corresponds to a reduction of about 40% in the emissions from road traffic.

28. Measured maximum hourly concentrations over the past five years are shown in Table II.7.7, from which it can be seen that the greatest concentrations measured in recent years lie in the range 352-423 ppb. Using the approach described above leads to the conclusion that to reduce this figure to 150 ppb (286.5 µg/m^3) would mean a reduction of between 40-48% in urban road transport NO$_x$ emissions relative to 1995 values.

Meeting Long Period Air Quality Objectives

29. The urban background locations at Bridge Place and London Bloomsbury had the highest annual average nitrogen dioxide concentrations of 34 ppb (65 µg/m^3) and (35 ppb (67 µg/m^3) respectively in 1995 as measured by the automated network. This value exceeds the WHO annual guideline of 20.9 ppb (40 µg/m^3). It can be shown by using monitoring data stretching back to 1976 that the annual average NO$_x$ concentration is linearly related to the annual average nitrogen dioxide concentration and is approximately 2.5 times greater. Thus, in order that these urban background locations do not exceed this guideline, reductions innitrogen dioxide concentrations of 40% are required. This then correlates with a required 40% reduction in urban emitted NO$_x$. If only road transport emissions are assumed to be reduced then a 48% reduction in NO$_x$ from this source would be required to attain the WHO annual guideline figure of 20.9 ppb (40 µg/m^3). Over the last 5 years the highest recorded annual average at an urban background location was 43 ppb at Bridge place in 1991. Using the same methodology as above this would lead to an emissions reduction requirement in urban road transport derived NO$_x$ of approximately 62%.

30. The highest background and intermediate site annual mean concentration of nitrogen dioxide from the Diffusion Tube Survey was 39 ppb at Hackney. Using the same procedure as outlined above a reduction in urban emitted NO$_x$ of at least 46% would be required. This equates to a 55% reduction from present day levels if the reduction comes only from road traffic sources.To summarise, it is apparent that reductions in urban road transport NO$_x$ emissions of between 48-62% on 1995 levels would be required in order to satisfy the WHO annual average guideline figure of 20.9 ppb in urban background locations.

31. In general the measured concentrations of nitrogen dioxide at roadside locations are governed by the available oxidant (ozone), except under very polluted conditions when another mechanism also brings about the conversion of nitric oxide to nitrogen dioxide. Measurements over many years have shown that an annual average concentration of nitrogen dioxide equal to 20.9 ppb corresponds to an annual average NO$_x$ concentration of approximately 52 ppb. Hence, NO$_x$ concentrations at roadside sites must be reduced to this level if the annual average guideline is to be attained. The measured NO$_x$ concentrations at Cromwell Road have been in the range of 205-273 ppb over the last 6 years where the

value for 1995 was 206 ppb. This suggests that reductions in urban NO_x of 74-81% would be required in order to reach the annual average guideline for nitrogen dioxide. The proportion of these measured levels from road transport is higher than in background locations and for the purposes of this analysis is taken as 100%. Hence, the corresponding reductions from urban road traffic would also be in the range 74-81%.

Conclusions 32. The Government's objective is to reduce ambient nitrogen dioxide levels to the extent that annual average levels are kept continuously low and peak episodes of wintertime smog are avoided.

33. The Government has decided to accept 150 ppb as an hourly mean as the standard for nitrogen dioxide, with the achievement of this value by 2005 as a provisional objective. The Government has also decided to adopt the further provisional objective of 21 ppb as an annual mean to be achieved by the year 2005. The UK will press for vehicle emission limits and fuel standards, to apply within thc EU from 2000, consistent with achieving those objectives by 2005. Tables II.7.7 and II.7.8 show the exceedences of the objectives for the period 1990 to 1995.

34. The Government has decided that the provisional objective for hourly nitrogen dioxide concentrations should apply at any non-occupational near-ground level outdoor locations. The Government has decided that the provisional annual average objective should apply in the following non-occupational outdoor locations: background locations; roadside locations; and other areas of elevated nitrogen dioxide concentrations where a person might reasonably be expected to be exposed (e.g. in the vicinity of housing, schools or hospitals etc) over the averaging time of the objective. Consideration of potential exceedences of the nitrogen dioxide provisional annual objective at potentially high concentration locations should be carried out in conjunction with data from urban or other appropriate background locations. This approach will be reviewed in the light of further research into patterns of personal exposure.

35. The analysis above shows that reductions in nitrogen oxide emissions from road transport of the order of 48-62% will be required on 1995 levels to enable attainment of the objectives in background urban locations in the UK. Reductions, perhaps in excess of 70%, may be required at roadside locations. Current policies should deliver reductions of about 38%. Forecasts suggest a reduction in urban road transport derived NO_x of 45% by 2005, on the basis of the European Commission proposals for post-2000 vehicle standards. With the possible provisional standards for post-2005 this reduction is 47%, though these standards will be subject to review by the Commission in 1998. This conclusion, and the approximate size of the further emission reductions required, are broadly consistent with the findings of the Auto-Oil Study which concluded that, for London, a reduction of 40% in NO_x emissions in 2010 relative to those estimated to result in that year from current policies.

36. It is worth noting that the emissions forecast shows that the trend of decreasing emissions is set to reverse some time after 2010 due to the expected growth in traffic. Therefore, considerations of sustainable development dictate that further control measures will be required to secure the downward trend in emissions. Possible considerations for measures involving new vehicle and fuel technologies might include some of the following:

- the benefits in emissions reductions from catalyst equipped cars can only be achieved when the catalyst is functioning efficiently, i.e. when it has fully warmed up. It has been shown that there is a cold start penalty for 3-way catalyst fitted cars. On a typical urban drive the cold start emissions of NO_x relative to those from a fully warmed up vehicle were about 30% higher. A conventional underbody catalyst presently takes up to 160 seconds to 'light off' (ie. the point when the catalyst is sufficiently operational so as to give a 50% reduction in hydrocarbon emissions) which corresponds to a journey length of about 3 km. Most journeys in the UK are short with about 40% being less than 5 km;

- close coupled catalysts have a light off time which is much shorter and of the order of 70 seconds. Thus 'cold start' emissions reductions of 50% in hydrocarbons and NO_x, and 18% in carbon monoxide are possible with such devices;

- the Auto-Oil Programme, and more specifically the EPEFE study (European Programme on Emissions, Fuels and Engine Technologies) has shown unequivocally that sulphur in gasoline reduces the efficiency of any optimally functioning three-way catalytic converter. Sulphur levels of the order of 300 ppm are found in UK gasoline. Reduction of this concentration to 30 ppm will reduce a car's post-catalyst (when warm) emissions of hydrocarbons by 50%, its NO_x by 20% and its carbon monoxide by 40%. Measures incorporating improvement in vehicle technologies and fuel specifications are anticipated in proposals from the EC in the near future. Bearing in mind the requirements for NO_x reductions to achieve the NO_2 levels discussed above and the objective for ozone, the UK will press for the most stringent standards for 2000 and 2005, subject to an assessment of costs and benefits; and

- it is apparent from the forecasts of urban NO_x emissions that the proportion of emissions from what can be termed ' urban fleets' (ie. vehicles which have high mileages in urban areas and are usually depot based) are set to increase in the future. For instance, at the present time emissions from LGVs, buses and small HGVs contribute 28% of all urban NO_x. This is predicted to rise to 38% in 2010. This type of vehicle is well suited to run on alternative fuels, such as Compressed Natural Gas and Liquefied Petroleum Gas, which offer substantial emissions benefits over diesel fuel. Increased gas fuel usage in urban delivery and public service vehicles could help reduce urban NO_2 concentrations.

37. A range of other possible non-technical measures acting on transport emissions is conceivable, which would not only reduce NO_x emissions, but also those of other pollutants as well, and these are discussed in the chapter devoted to transport issues. The policy gaps implied by the various objectives are discussed in Part I, as are the uncertainties of meeting the reductions in levels of pollutants in all locations in the UK.

Chapter II.8: Ozone

Introduction and Sources

1. Ozone is not emitted directly from any man-made source in any significant quantities, but arises from chemical reactions in the atmosphere caused by sunlight. In the stratosphere, where ozone plays a beneficial role by shielding the earth from harmful ultra-violet radiation, ozone is produced by sunlight acting initially on oxygen molecules. The balance between ozone and oxygen is currently being disturbed by migration upwards of chemicals such as chlorofluorocarbons, which remove ozone and may therefore increase the amount of ultra-violet light reaching the earth's surface.

2. In the lower layers of the atmosphere, while some ozone occasionally arises from periodic intrusions of air from the stratosphere, it is primarily formed by a complicated series of chemical reactions initiated by sunlight. In these, oxides of nitrogen (NO_x) and hydrocarbons (or VOCs - volatile organic compounds), derived mainly from man-made sources, react to form ozone. These substances are produced by combustion, other industrial processes, and other activities such as solvent use, and petrol distribution and handling (see Table II.7.1 and Table II.8.1). Although NO_x and VOCs are the most important precursors of elevated levels of ozone, production of ozone can also be stimulated by carbon monoxide, methane, or other VOCs which arise from plants, trees and other natural sources. Ozone, NO_x, and VOCs are also indirect greenhouse gases which influence atmospheric chemistry cycles and hence atmospheric radiative forcing.

3. These chemical reactions do not take place instantaneously, but over several hours or even days depending on the VOCs, and once ozone has been produced it may persist for several days. In consequence, ozone measured at a particular location may have arisen from VOC and NO_x emissions many hundreds, or even thousands of kilometres away, and may then travel further for similar distances. Maximum concentrations, therefore, generally occur downwind of the source areas of the precursor pollutant emissions. Indeed, in urban areas, where concentrations of traffic gases may be high, nitric oxide (NO) from exhaust emissions may react with ozone to form nitrogen dioxide (NO_2) reducing ozone concentrations. However, as the air movement carries the primary pollutants away, more ozone is generated and concentrations rise in the downwind areas.

4. In terms of ozone measured at ground level, these photochemical episodes of high ozone concentrations are superimposed on a baseline which varies slightly throughout the year but averages around 30 ppb at UK latitudes. This is made up partly of ozone transported from the stratosphere, and some ozone produced in the troposphere (the region of the atmosphere, about 10 km deep, between the Earth's surface and the stratosphere) from naturally occurring and man-made precursors (in broadly equal proportions). This is shown schematically in Figure II.8.1. There is evidence that this baseline has roughly doubled since the turn of the century, largely due to the increase in man-made NOx emissions arising in the whole of the Northern Hemisphere over this period. As will be discussed below, the magnitude of this baseline is close to levels at which effects have been observed on crops and plants.

5. From these considerations, and bearing in mind the importance of sunlight in the reactions, elevated ozone levels occur more frequently in summer, in the southern UK more than the north and in rural and

suburban areas more than in city centres. Moreover, in northwest Europe, because of the time taken for ozone to form and then be destroyed in the atmosphere, and hence the distance it can travel, the problem is an international one. Unilateral action by any one country would be of limited effectiveness in the overall reduction of ozone levels.

Health Effects:
Standards and
Guidelines

6. Ozone toxicity occurs in a continuum in which higher concentrations, longer exposures and greater activity levels during exposure cause greater effects. Short-term effects include pulmonary function changes, increased airway responsiveness to broncho-constrictors, and airway inflammation. These health effects are generally significant at exposures as low as 80 ppb for 6.6 hours in a group of healthy exercising adults. Controlled exposures of heavily exercising adults or children to 120 ppb for 2 hours also produce decrements in pulmonary function. Substantial acute adverse effects are clear at concentrations of 250 ppb or higher, particularly in susceptible individuals or subgroups. There is also evidence that ozone levels of about 120 ppb can cause airway changes which increase the sensitivity of subjects to inhaled allergens such as pollen. To date there is no evidence that the adverse effects due to ozone cause long-term damage to the respiratory system at the levels likely to occur in the UK. Nor is there evidence to suggest that asthmatics are significantly more sensitive to ozone than other members of the population.

7. Taking this evidence into account, and noting that the ozone exposure studies did not use the more sensitive members of the population, EPAQS recommended an air quality standard for ozone of 50 ppb, expressed as a running 8-hour mean.

8. As a parallel exercise, within the UNECE framework, critical levels are being developed for the impacts of ozone on semi-natural ecosystems, crops, plants and trees. Although no guidelines, standards or critical levels have been formulated for the effects on materials, ozone damage, particularly to rubbers and paints, is also well known.

Current Air
Quality

9. Concentrations in the UK have been monitored since 1972 at a number of sites, largely in the southern part of the UK. These measurements were often sporadic and in the mid 1980s the Department of the Environment sought to co-ordinate ozone monitoring in the UK. The Photochemical Oxidants Review Group (PORG) was set up to advise on the planning of the network, and by 1987 18 sites were established and producing data. In 1992 the existing information base, consisting largely of rural sites, was supplemented by the ozone monitoring in the Automatic Urban Network, giving as of January 1997, a total of 53 sites in the DoE networks, with a total of 74 sites expected by the end of 1997.

10. Peak ozone concentrations are very dependent on meteorological conditions in that they require not only low wind speeds to build up high levels of precursor pollutants and strong sunlight to promote the chemical reactions, but also air mass trajectories which pass over precursor emission source areas. A typical series of trajectories leading to an ozone episode in the UK in 1995 is shown in Figure II.8.2, from which it is clear that emissions from Europe and the UK have been picked up and have contributed to the levels of ozone measured at various locations in the UK.

11. The magnitude and frequency of high ozone episodes can therefore vary considerably from year to year and this makes analysis of trends difficult. There is nonetheless evidence beginning to emerge that over the period since the mid-1970s there has been a downward trend in the highest hourly ozone values observed in each year. The highest values ever observed in the UK were in the summer of 1976 when the UK and northwest Europe experienced a very long period of hot, dry, still and sunny weather. The highest hourly average recorded was 258 ppb at Harwell, and levels were probably higher than this as the recorded value was at the maximum of the measurement range in use in the instrument at the time. In the last decade the highest hourly values measured have been in the range 100-150 ppb.

12. It is not possible to be confident about the reasons for such a decrease since there are very few measurements of precursor pollutants in the UK or Europe over this timescale. There has nonetheless been a gradual tightening of motor vehicle emission limits over this period, and there is published evidence from an analysis of measurements made on the west coast of Ireland that this has resulted in a downward trend in carbon monoxide emissions from Europe over the past 8 years or so. It is reasonable to suppose therefore that, for similar reasons, there may well have also been a decrease in VOC emissions, which overall are more potent ozone producers than carbon monoxide. However, this must be set against the variability in the weather from year to year, where as has been noted above, strong sunlight and high temperatures alone are not sufficient to generate high ozone levels without the 'favourable' air-mass trajectories picking up precursor emissions from elsewhere in Europe. An initial assessment of the meteorological conditions in 1976 and the hot summer of 1995 suggests that differences in the tracks of the trajectories could have caused some of the differences in ozone levels. A more detailed analysis of ozone trends in the UK is being undertaken by the Photochemical Oxidants Review Group and will be published later this year.

13. As has already been discussed above, elevated ozone concentrations are higher and more frequent in the southern parts of the UK. Moreover on higher ground, which is often above any shallow overnight temperature inversion within which ozone is destroyed by dry deposition or NOx, elevated ozone levels can persist longer than in low lying areas, so that a map of the frequency of occurrence of elevated ozone levels in the UK has the form shown in Figure II.8.3

14. The EPAQS recommended standard for ozone has been exceeded in at least one year at virtually every site for which ozone measurements are available. During 1990, when photochemical ozone episodes were widely prevalent in the British Isles, the EPAQS recommended standard was exceeded at all the UK monitoring sites from 7 to 83 days at each site as shown in Table II.8.2. The frequency of exceedence of the standard is generally highest at the monitoring sites in the south and east of the British Isles and least in the north and west.

15. Figure II.8.3, which maps the number of hours above 60 ppb ozone, should display quite accurately the likely exceedence of the EPAQS standard, since 60 ppb as an hourly mean is, on average at UK sites, equivalent to the 8 hour value of 50 ppb recommended by EPAQS. The map shows a relatively low likely exceedence of the EPAQS ozone

standard in the Highlands of Scotland and Northern Ireland with increasing exceedence across England and Wales moving southwards and to a slightly lesser extent eastwards. The populated regions of southern and southeast England, and the upland regions of the Pennines, Wales and the southwest stand out with increased exceedences of the EPAQS recommended standard.

16. This spatial pattern of likely exceedence of the standard reflects:

■ the more frequent occurrence of summertime anticyclonic sunny weather in the south and east;

■ the importance of long range transport from the more dense precursor emissions of northwest Europe compared with the regions to their north or south;

■ the importance of altitude and of local nocturnal atmospheric stability on the diurnal pattern of photochemical ozone in rural locations.

17. Table II.8.3 presents data on the maximum 8-hour running mean ozone concentrations recorded for a selection of sites in the 1990's based on ozone monitoring network data. Peak 8-hour mean ozone concentrations throughout the UK appear to exceed 100 ppb.

18. Two factors can therefore be quantified from the available monitoring data concerning the exceedence of the EPAQS recommendation, namely, the frequency of exceedence (see Table II.8.2) and the extent of exceedence (see Table II.8.3). Both these factors show significant year-on-year variability which needs to be taken into account in considering these exceedence data. Consideration of the concentrations in Table II.8.3 suggests that maximum 8-hourly levels of ozone would need to be reduced by about 55% to attain the EPAQS recommended value.

19. Ozone monitoring in the EU is governed by the 'Ozone Directive' (92/72/EEC). Rather than setting mandatory limit values, this Directive embodies thresholds related to health effects at which information and warnings have to be given to the public. These thresholds are 90 ppb (180 µg/m3) and 180 ppb (360 µg/m3), respectively, as hourly means. The Warning Threshold has not been exceeded since 1981 (prior to the Directive), but the Information Threshold has been exceeded in every year in which monitoring has taken place. The magnitude of these exceedences is shown in Table II.8.4. Inspection of the values measured in the last 10 years suggests that the maximum hourly values would need to be reduced by about 35-45% to achieve levels below the EC Information Threshold.

The Strategy 20. Unlike the so-called primary pollutants, such as benzene, whose ambient air concentrations are directly proportional to the quantity emitted, the chemical reactions by which ozone is formed are complex and ozone levels in one location will not, in general, bear a simple relation to emissions of VOCs and NO_x at that location. Since the chemical reactions involved can take from hours to days to occur, ozone levels can be influenced by emissions from many hundreds, even thousands, of kilometres distant. Ozone levels in the UK, therefore, are influenced not only by national emissions of VOCs and NO_x, but by emissions from northwest Europe and elsewhere, so that unilateral action by any one country is unlikely to make a significant impact on UK ozone levels.

21. Moreover, a further feature of the atmospheric processes which form

ozone is the fact that, while it appears that reductions in VOCs always lead to reductions in ozone levels, reductions in NO_x emissions could lead to increases in ozone in some areas. This is due to the complexity of the atmospheric chemistry where, near to emission sources, NO_x can destroy ozone even though it contributes to its formation further downwind. Recent studies have identified broad regions of Europe where the atmosphere's capacity to produce ozone (which is large) is limited by the availability of (a) VOCs and (b) NO_x, and these are shown schematically in Figure II.8.4. More detailed analysis suggests that hydrocarbon limited regions are found in the most densely populated and industrialised regions of northwest Europe and in isolated urban industrial centres on the Mediterranean coast. Much of the densely populated areas of the south and east British Isles are hydrocarbon limited regions. In these regions, ozone levels will respond more effectively to hydrocarbon emission controls or to some combination of both NO_x and hydrocarbon control.

22. In the rest of Europe, the amount of NO_x available is limiting and ozone levels are determined by the relative efficiency of long range ozone transport from the hydrocarbon limited regions and ozone dry deposition. In these regions, more remote from the major source areas, local hydrocarbon emission control is usually ineffective whilst NO_x control is highly effective.

23. This understanding of the mechanisms of ozone production in Europe has allowed the development of numerical models, which can be used to estimate the effects of current policies on ozone concentrations and the extent to which further reductions may be required to reduce concentrations to achieve air quality standards and objectives.

24. The concept of hydrocarbon and NO_x limited regions can be explored in more detail using such numerical models and Figure II.8.5 shows the results of calculations using the UK Meteorological Office (UKMO) photochemical trajectory model. These calculations have used air mass trajectories for actual ozone episodes in the UK and have in turn reduced emissions of hydrocarbons, NO_x, and both hydrocarbons and NO_x by 50% in both the UK and Europe and compared the resulting ozone concentrations with the no-reduction case. The results in Figure II.8.5 show which of the three emission reduction scenarios is most effective in reducing peak ozone concentrations. To address ozone levels in the southern half of the UK, a strategy involving reductions of hydrocarbon (VOCs) emissions, or of both VOCs and NO_x appears to be most effective in reducing peak ozone. The areas delineated in Figure II.8.5 are approximate and illustrative and should not be compared in detail with those in Figure II.8.4.

25. Against this background, an important question is whether action in the UK alone would be effective in controlling ozone concentrations, or whether episodes would still occur because of long range transport of ozone from sources in the rest of Europe.

26. A second question is whether UK emissions would themselves cause ozone episodes if other European emissions were eliminated. A third important question is whether the regional scale ozone problem is in any sense controllable. These questions have been addressed using the UKMO model referred to above. The results are summarised in Table II.8.5 and the conclusions are as follows.

27. Action taken unilaterally in the UK is likely to have a significant influence on the extreme peak ozone concentrations in the UK (column 3 of Table II.8.5). Ozone episodes would still occur if UK emissions were eliminated, albeit with reduced spatial coverage and severity, due to long range transport. Elimination of UK emissions is unlikely to change exceedences of the EPAQS recommended level to any significant extent, but is likely to reduce exceedences of the EC Ozone Directive Information Threshold.

28. With UK precursor emissions alone and sources in the rest of Europe eliminated (column 4 of Table II.8.5), ozone episodes would still occur, but with a very much reduced coverage and severity. The calculations confirm the importance of European sources in influencing the spatial coverage and intensity of ozone episodes over the UK. Whatever measure of ozone exposure is taken, precursor sources in the rest of Europe appear to be at least twice as important as those within the UK. Action to control precursor emissions in the rest of Europe is a vital prerequisite to reducing exceedences in the UK of the EC Ozone Directive Information Threshold and the EPAQS recommended level.

29. If precursor emissions due to human activities in the whole of Europe are set to zero, no photochemical ozone production is apparent. The model includes natural emissions of VOCs, e.g. from trees, but these make very little contribution to ozone episodes occurring in the UK. In principle, therefore, emission reductions of sufficient severity could achieve both the EC Information Threshold and the EPAQS recommended value, but considerations of abatement costs and benefits will be important in determining the extent to which these levels are achieved. Table II.8.1 summarises the sources of VOCs in the UK.

30. Following the broad strategic issues addressed above, it is important to assess the extent to which current policies will affect ozone concentrations.

31. Policy instruments for the control of ozone precursors are in place at a national and European level. At the European level, the UK is committed to action through two fora, the United Nations Economic Commission for Europe (UNECE) and the European Community.

32. In 1979 the UNECE adopted a Convention on Long Range Transboundary Air Pollution (LRTAP) to which the UK is a party. It has been followed by a series of Protocols (on sulphur dioxide, NO_x and VOCs), setting out specific targets for action. In November 1991, the UK signed the VOCs (Geneva) Protocol, and ratified it in June 1994. The Protocol primarily requires a reduction in VOCs emissions of 30% by 1999 compared to 1988 levels. Other obligations include: the application of national or international emission standards to new stationary and mobile sources; the application of national or international measures to products that contain solvents; the promotion of the use of products with low or nil VOCs content; the fostering of public participation in VOCs emission control programmes through public announcements; the encouragement of the best use of all modes of transport; and the promotion of traffic management schemes.

33. Current projected emissions suggest that the target will be more than met (see Table II.8.1). However, the Protocol has not as yet been ratified by all EC signatories so that the extent of VOCs emission reductions in the EC and elsewhere in Europe are unclear. At present therefore, reductions

in VOC emissions of around 30% from Europe as a whole appear to be the best approximate estimate, given that some signatories to the Protocol (the UK included) may reduce by more than 30% figure.

34. The UNECE NO$_x$ Protocol was agreed in 1988 in Sofia and committed signatories to return their 1994 NO$_x$ emissions to 1987 levels and to maintain them below that level subsequently. The NO$_x$ Protocol is currently being revised within the UNECE and given the central role played by NO$_x$ not only in ozone formation but also in ecosystem acidification and eutrophication, the revised Protocol is "multi-pollutant" and "multi-effect" in nature in that all the above effects are being considered as are options to abate not only NO$_x$, but VOCs, ammonia and also possibly sulphur dioxide, if this is cost- effective.

35. The EC (which is a party to LRTAP) has adopted a series of Directives tackling, inter alia, VOCs emissions from mobile sources:

■ 'EURO I' vehicle emissions Directives which set limits for the emissions of nitrogen oxides, hydrocarbons, carbon monoxide and particulates (diesel vehicles only). In practice EURO I required the fitting of catalytic converters to all new petrol cars;

■ Directive 93/441/EEC, passenger cars (from 31/12/1992);

■ Directive 93/59/EC, light vans (from 1/10/1994);

■ Directive 91/542/EEC, heavy duty vehicles (from 1/10/1993);

■ 'EURO II' vehicle emissions Directives which tighten further the limits in EURO I;

■ Directive 94/12/EC, passenger cars (from 1/1/1997);

■ Directive 96/69/EC, light vans (from 1/10/1997);

■ Directive 91/542/EEC, heavy duty vehicles (from 1/10/1996);

■ benzene content of petrol is controlled (Directive 85/210/EEC).

36. Furthermore, in assessing the effects of current policies, it has been assumed that the Stage I Petrol Vapour Recovery Directive (94/63/EEC) will be implemented, in 1996 on new installations and in 1999 on existing facilities, and that diesel car sales would level off at 20% of all car sales from 1994.

37. Directives further tightening vehicle emissions and fuel standards and on solvents are anticipated. The Directive on Stage II Petrol Vapour Recovery now appears unlikely to emerge from the EC, and consideration is being given to implementation of such measures in the UK. However, it is clearly too early to incorporate any effects of these measures at this stage.

38. National legislation is directed at meeting both international commitments and local pollution concerns. As described in Chapter 5, in 1990 the UK introduced a new pollution control regime under the Environmental Protection Act 1990 (EPA 90). This subjects a range of industrial processes, some of which emit VOCs, to new tighter controls.

39. As noted above, current projections suggest that VOCs emissions in 1999 could reduce by 37% compared with the 1988 base on the basis of the application of EPA 90. Action under EPA 90 will also lead to reductions in NO$_x$ emissions from stationary sources and current indications are that the targets for 2003 in the Large Combustion Plant Directive will be met (namely reductions of 30% by 1998 on a 1980 baseline). The measures put in place to achieve these targets coupled with further applications of

EPA 90 will lead to more reductions in stationary source NO_x.

40. The foregoing discussion has shown that current policies are not sufficient to achieve ozone guidelines related to human health effects (see Table II.8.3). The question then remains as to the extent of further abatement which might be required to attain such levels. This has been addressed using the UKMO trajectory model applied to series of six ozone episodes which occurred in 1993 and 1994. These calculations used actual air mass trajectories and emission inventories for the UK and the rest of Europe. Following the simulation of each episode, model calculations were carried out reducing emissions of VOCs and NO_x by 25%, 50%, 75% and 90% in both the UK and the rest of Europe in each episode in turn.

41. Such calculations generate a considerable amount of information (giving ozone concentrations for each 100 by 100 kilometre OS grid square in the UK in each calculation). It is therefore necessary to summarise the results using some indices of ozone concentration. This has been done in Table II.8.7 and Figure II.8.6.

42. The term 'maximum' refers to the maximum ozone concentration recorded over any of the six days and 50 arrival points. The other terms refer to the mean ozone concentrations calculated to reach each arrival point over the six days. 'Most exposed' refers to the ranked mean ozone concentrations and the concentration experienced by the most exposed 0.5 million population. 'Population weighted mean' or 'mean exposed' refers to the population weighted mean ozone concentration. 'Area weighted mean' refers to the mean ozone concentration across southern England.

43. The entries plotted in Figure II.8.6 are shown as parallel lines, 20 ppb apart. The base case model formulation includes a tropospheric ozone background of 50 ppb at the start of the air parcel trajectory. A comparison of model results with the rural ozone network data for the six chosen days shows that this background may be too high on occasions. On these latter days a value closer to 30 ppb would be more appropriate. Unfortunately, European ozone monitoring networks are not developed sufficiently in terms of spatial and temporal distribution to provide a better model initialisation. The parallel curves are accordingly used to illustrate the impact of the uncertainty due to model initialisation. Clearly, the initial ozone concentrations should in fact be scenario dependent but this dependence has been neglected here also.

44. Maximum ozone concentrations fall below the 90 ppb level with percentage reductions beyond 60-80%. Such percentage reductions may be at the limit of what is currently considered technically feasible albeit at large costs.

45. The concentrations experienced by the most exposed 0.5 million of the UK population fall below the 90 ppb level with percentage reductions of about 30% and below the 60 ppb level with percentage reductions beyond 55-85%. Population weighted mean ozone concentrations fall below the 60 ppb level with percentage reductions beyond 75%.

46. From these results it appears that abatement strategies aimed at reductions in VOCs and NO_x emissions in the region of 50-70% from the present day will make significant reductions in the maximum ozone concentrations experienced in the UK and in the extent of the population at risk of exposure to elevated ozone concentrations in excess of health related thresholds.

47. It should be repeated here that further UK action will need to be part of a coherent European strategy which will need to take full account of the costs and benefits of abatement. We have seen how, because of the disparity in possible actions to reduce VOCs across Europe, there is the risk that emissions of VOCs from the rest of Europe will not fall as quickly as those from the UK. Since peak ozone levels in those parts of the UK where they are highest, will respond more effectively to reductions in VOC emissions both from the UK and from the rest of Europe, the UK will press for European action to achieve further substantial reductions in VOC emissions. This will involve pressing for EC-wide ratification of the UNECE VOC Protocol as part of the EC Ozone Strategy, and for cost-effective VOCs abatement in the UNECE.

48. Development of such strategies will not be easy as in many countries in northwest Europe at least, many of the less costly measures are already in place or planned. Strategies will therefore need to be optimised in order to obtain the most cost-effective solutions. Indeed, strategy optimisation was one of the successful lessons learned in the negotiation and formulation of the second UNECE Sulphur Protocol. There are several options to explore in devising an optimised strategy for Europe. Reduction in emissions should be concentrated on those areas which contribute most to the environmental problems at issue. This was the basis behind the UNECE Sulphur Protocol and, despite the additional complexities, is likely to be the basis for the revised NO_x Protocol currently under negotiation. The UK fully supports this approach and is contributing positively to the development of the Integrated Assessment Models required to carry it out.

49. In considering optimal strategies for ozone, it should be possible to exploit further the seasonal dimension of the ozone problem, and measures which do so are likely to provide better value for money. On a similar theme, while substantial reductions are required to achieve the thresholds discussed above, meaning that long term measures are necessary, it will also be important to consider the effectiveness of short term measures, involving traffic management or other interventions (for example in the industrial sector), in reducing peak ozone levels.

50. Furthermore, to date all UK analyses of abatement costs for VOCs have reported results in terms of the cost (or marginal cost) per tonne of VOCs abated in the context of ozone. This does not optimise the widely different potential for producing ozone exhibited by the range of VOCs. This important property of VOCs was stressed by the UK in the development of the UNECE VOCs Protocol, which contains an exhortation to signatories to take the so-called Photochemical Ozone Creation Potentials (POCPs) into account in devising strategies. Now, however, with abatement cost curves for VOCs beginning to emerge in the UNECE forum, and emission inventories developing (in the UK and elsewhere) which are desegregated (or speciated) into individual VOCs - the current UK inventory now contains some 300 individual VOCs - it should be possible to incorporate POCPs directly into the modelling of optimal abatement strategies for ozone. The important quantity in this case being not the cost per tonne of VOCs abated but the cost per tonne of ozone formation avoided.

51. A further concept, which becomes more important as the less costly options are taken up, is the principle that the burdens should be equalised

as much as possible across the various industrial sectors by selecting, to the extent possible, abatement options with comparable marginal costs of abatement (incorporating POCPs where feasible).

52. As has already been noted above, the current plans for reductions in UK VOCs emissions should result in a decrease of some 37% by 1999 compared with 1988 levels. A recently published study of abatement technologies and costs carried out on behalf of the DTI suggests that reductions of the order of 52% are likely to be the maximum achievable on the basis of existing available technologies for which reliable costs could be obtained (although the study suggests that, allowing for the development and increased adoption of new technologies, further reductions of around 5% could be obtained in the sectors examined). The average cost quoted for this 'maximum cut' scenario is about £8,000/tonne of VOCs abated (£8,400/tonne is quoted in the report) compared with £660/tonne for the 'existing commitments' scenario (delivering the 38% reduction). However, a 'least cost' version of the maximum cut scenario was also investigated and this lead to average costs of £1,200/tonne.

53. Within the EC, discussions are currently under way on a further series of limits for motor vehicle exhausts and for petrol and diesel fuel quality. Both measures could contribute significantly to reducing ozone concentrations in Europe. They will also be beneficial in reducing a number of other environmental problems involving urban nitrogen dioxide, particles, carcinogenic VOCs, carbon monoxide and acidification. One of the important conclusions of the Auto-Oil study, which informed the formulation of the Commission's proposals in this area, was that the reduction of ozone concentrations will require action on stationary as well as mobile sources. The UK will press for substantial reductions in ozone precursor emission from mobile sources and fuel standards in the discussions on these proposals.

54. A major area of stationary source VOCs emission in the UK and the rest of Europe is the use of solvents, in painting and coating, printing etc. and a Directive aimed at reducing VOCs emissions from this sector is in preparation within the EC. The DTI study mentioned above suggests that significant reductions are achievable, of the order of 14% reduction on 1988 levels, at an average cost of about £1400/tonne of VOCs abated.

55. Current policies are expected to reduce UK NO_x emissions by about 50% by 2010 (see Table II.8.6). Further reductions should accrue by this date from the EC proposals on vehicle emissions and fuels. The other major source sector comprises the large combustion plants - the electricity supply industry, oil refineries and other large industrial plant. Emissions from this sector are controlled by EPA 90 and conform to the National Plan to deliver the Large Combustion Plant Directive requirements for NO_x emissions (namely reductions of 30% by 1998 on a 1980 baseline). NO_x emissions will further reduce in the current round of authorisations under EPA 90 as a result of the increased displacement of coal by gas in electricity generation and the use of low-NO_x burners in coal-fired plants. Further reductions could require the widespread use of low-NO_x burners and possibly the installation of Selective Catalytic Reduction in coal-fired plant.

Conclusions

56. The reduction of ozone concentrations is difficult and requires a coherent European approach to achieve the best value for money in expenditure on abatement measures. While there is merit in the UK acting alone in that some reductions in peak levels will occur, maximum benefit will be obtained by a concerted approach with our neighbours. The location of the areas of the UK which experience the highest peak ozone levels in relation to the emission sources in the UK and the rest of Europe means that control of VOCs alone, or in conjunction with NO_x will be most effective in reducing peak ozone levels in the UK. Emissions elsewhere in Europe appear to be roughly twice as important as those within the UK for reducing peak ozone levels.

57. The Government recognises that the level recommended by EPAQS, while achievable in principle by reducing man-made emissions, would require extremely large reductions. Nonetheless, the Government has decided to accept 50 ppb as a running 8-hour mean, as the air quality standard, defining the level of ozone at which effects on public health, including the most sensitive individuals, would be small. The Government further notes the comment of EPAQS that restricting the number of exceedences of the standard to 10 days or so per year at any one site would ensure that the maximum 8-hour values at that site would be unlikely to exceed 100 ppb, a concentration at which effects in healthy individuals have been clearly demonstrated.

58. Reductions of VOC and NO_x emissions in the range 50-70% relative to the present day, while they would not completely remove exceedences of either the EPAQS level or the EC Ozone Directive information threshold, would nonetheless significantly reduce the extent to which they are exceeded. Exceedences of the EPAQS value would be eliminated over most of Scotland and would be significantly reduced over the rest of the UK. Exceedences of the EC Information Threshold would be largely eliminated over substantial areas of Northern Britain.

59. The Government has decided to adopt 50 ppb as a running 8-hour mean as a provisional objective for ozone, to be achieved[1] at the 97th percentile level, i.e. on all but 10 days per year assuming perfect operation of the monitoring station with 100% data capture throughout the year, by 2005. The Government will move towards this objective as quickly as costs and benefits allow, with the intention of securing in effect, substantial elimination of summertime smog episodes, defined as exceedences of the EC Directive Information Threshold.

60. The Government has decided that the provisional objective for ozone should apply in all near-ground level non-occupational outdoor locations where a person might reasonably be expected to be exposed over the averaging time of the objective. The provisional objective will be pursued within the overall aim of arriving at a coherent and cost-effective ozone strategy for Europe taking into account the effects of ozone precursors such as NO_x, on acidification, eutrophication, urban nitrogen dioxide levels and other appropriate effects.

---

[1] This could be expressed as a percentile, defined for simplicity in the following way, to avoid the confusion of multiple running means. The maximum running 8-hour mean in any one day is assigned to that day, so that a series of 365 8-hour means is generated. Using the method of calculating percentiles in the current EC Directive on NO_2 , assuming perfect operation of the monitoring station throughout the year, the target of the 11th highest day is, strictly, the 97.3 percentile, which could be rounded to the 97th for clarity. This is less stringent however, and it may therefore be preferable to retain the '10 days exceedence' criterion.

**Table II.8.1
Emissions of VOCs in
the UK, 1988-1999
(excluding forests)**

	Emissions (kilotonnes)*												% Reduction**
	1988	1989	1990	1991	1992	1993	1994	1995	1996	1997	1998	1999	
Industrial Solvent Use	572	591	576	540	512	515	509	503	434	397	306	299	48
Domestic Solvent Use	195	189	185	189	187	189	197	197	193	191	190	190	3
Oil Industry	391	369	371	373	378	388	412	404	430	426	382	335	14
Chemical Industry	152	152	152	152	152	152	152	152	137	125	61	63	59
Stationary combustion	89	81	72	74	70	69	59	52	51	51	51	50	44
Other Industry	79	81	82	83	83	83	84	84	85	85	86	87	-10
Other Stationary sources	85	85	81	77	71	55	55	54	54	54	54	54	36
Road transport	951	1004	978	967	912	836	761	690	610	558	512	472	50
Other transport	129	128	125	128	129	126	126	121	121	120	120	119	7
Total	2642	2679	2623	2582	2493	2413	2534	2257	2116	2008	1762	1670	37

* Figures rounded to nearest kilotonne
** Figures rounded to nearest 1%

Chapter II.8: Ozone

Table II.8.2:
The number of days
on which the ozone
concentration of
50ppb as an 8-hour
running mean was
exceeded in 1990-1996

Site	1990	1991	1992	1993	1994	1995	1996*
Rural							
Stevenage	23	11	21	10	–	–	–
Sibton	44	39	47	46	48	35	32
Aston Hill	56	31	36	26	39	46	29
Lullington Heath	83	41	53	47	57	53	45
Strath Vaich	27	37	26	15	30	22	16
High Muffles	28	39	39	24	28	29	29
Lough Navar	21	14	20	5	8	20	8
Yarner Wood	55	60	42	26	31	45	36
Ladybower	32	15	13	13	18	24	22
Harwell	43	42	46	20	38	41	29
Bottesford	20	19	12	5	13	32	27
Bush	17	3	15	0	8	6	15
Eskdalemuir	27	17	29	8	14	26	12
Great Dun Fell	35	25	44	16	26	36	3
Wharleycroft	25	41	30	20	20	38	–
Glazebury	17	16	21	9	10	20	14
Rochester	–	–	–	–	–	–	32
Somerton	–	–	–	–	–	–	44
London Teddington	–	–	–	–	–	–	3
Urban							
Bridge Place	7	0	4	6	14	0	7
London B'bury	–	–	9	4	5	10	5
London Bexley	–	–	–	–	14	27	16
Edinburgh Centre	–	–	–	0	0	2	3
Cardiff Centre	–	–	18	11	10	30	19
Belfast Centre	–	–	2	0	2	12	30
Birmingham Centre	–	–	5	2	8	16	15
Birmingham East	–	–	–	–	15	26	21
Leeds Centre	–	–	–	2	3	12	7
Newcastle Centre	–	–	1	0	–	–	5
Bristol Centre	–	–	–	3	6	19	11
Liverpool Centre	–	–	–	–	2	16	8
Hull Centre	–	–	–	–	4	7	10
Leicester Centre	–	–	–	–	12	26	16
Southampton Centre	–	–	–	–	12	5	3
Swansea	–	–	–	–	–	29	14
Middlesbrough	–	–	–	–	–	6	12
Wolverhampton Centre	–	–	–	–	–	–	17
London Brent	–	–	–	–	–	–	24
London Haringey	–	–	–	–	–	–	14
London Kensington and Chelsea	–	–	–	–	–	–	15
London Eltham	–	–	–	–	–	–	16
Manchester Piccadilly	–	–	–	–	–	–	8
Sheffield Centre	–	–	–	–	–	–	11
London Sutton	–	–	–	–	–	–	16
London Wandsworth	–	–	–	–	–	–	7

★ 1996 data provisional

**Table II.8.3:
Maximum running
8-hour mean ozone
concentrations reported
for 1990- 1996 (ppb)**

Site	1990	1991	1992	1993	1994	1995	1996*
Rural							
Stevenage	114	75	82	103	–	–	
Sibton	127	75	109	106	110	96	115
Aston Hill	115	82	92	86	89	93	89
Lullington Heath	143	98	96	98	97	126	85
Strath Vaich	73	69	26	62	70	84	73
High Muffles	101	76	91	102	87	90	95
Lough Navar	85	79	70	56	66	74	60
Yarner Wood	121	112	103	88	102	111	82
Ladybower	105	74	97	80	84	86	99
Harwell	111	82	104	89	100	97	90
Bottesford	118	73	75	78	79	87	102
Bush	93	71	71	49	69	77	82
Eskdalemuir	90	67	72	99	69	88	80
Great Dun Fell	101	82	117	104	79	90	57
Wharleycroft	90	73	99	104	75	98	–
Glazebury	82	65	100	76	64	83	82
Rochester	–	–	–	–	–	–	86
Somerton	–	–	–	–	–	–	84
London Teddington	–	–	–	–	–	–	61
Urban							
Bridge Place	94	49	57	79	86	40	78
London Bloomsbury	–	–	65	81	73	85	74
London Bexley	–	–	–	–	88	102	81

★ 1996 data are provisional

Table II.8.3: Continued

Site	1990	1991	1992	1993	1994	1995	1996★
Edinburgh Centre	–	–	–	38	48	64	60
Cardiff Centre	–	–	93	77	94	95	80
Belfast Centre	–	–	54	44	53	68	65
Birmingham Centre	–	–	69	68	58	82	86
Birmingham East	–	–	–	–	90	83	94
Leeds Centre	–	–	–	50	62	62	75
Newcastle Centre	–	–	50	43	–	–	72
Bristol Centre	–	–	–	74	61	82	73
Liverpool	–	–	–	42	53	73	70
Hull Centre	–	–	–	–	87	69	73
Leicester Centre	–	–	–	–	80	85	87
Southampton Centre	–	–	–	–	77	85	65
Swansea	–	–	–	–	–	92	69
Middlesbrough	–	–	–	–	–	83	87
Wolverhampton Centre	–	–	–	–	–	–	93
London Brent	–	–	–	–	–	–	92
London Haringey	–	–	–	–	–	–	82
London Kensington and Chelsea	–	–	–	–	–	–	82
London Eltham	–	–	–	–	–	–	91
Manchester Piccadilly	–	–	–	–	–	–	75
Sheffield	–	–	–	–	–	–	87
London Sutton	–	–	–	–	–	–	83
London Wandsworth	–	–	–	–	–	–	72

★ 1996 data are provisional

**Table II.8.4:
Maximum 1-hour mean
ozone concentrations
reported for
1990-1996 (ppb)**

Site	1990	1991	1992	1993	1994	1995	1996★
Rural							
Stevenage	136	84	89	133	–	–	–
Sibton	145	90	123	122	127	112	121
Aston Hill	124	87	101	96	93	100	104
Lullington Heath	161	124	104	117	117	134	91
Straith Vaich	80	74	77	66	72	86	74
High Muffles	111	90	100	114	96	97	105
Lough Navar	94	94	86	62	70	88	68
Yarner Wood	147	126	115	99	108	130	95
Ladybower	120	86	114	101	112	97	104
Harwell	132	96	123	96	114	116	105
Bottesford	138	92	95	97	83	99	106
Bush	100	81	85	60	71	79	90
Eskdalemuir	106	81	80	115	75	92	86
Great Dun Fell	114	94	141	119	87	94	61
Wharleycroft	102	95	107	115	87	104	–
Glazebury	96	81	113	107	71	100	90
Rochester	–	–	–	–	–	–	99
Somerton	–	–	–	–	–	–	88
London Teddington	–	–	–	–	–	–	65
Urban							
Bridge Place	108	63	76	98	102	56	90
London Bloomsbury	–	–	73	93	95	102	90
Edinburgh Centre	–	–	–	44	55	71	72
Cardiff Centre	–	–	98	85	107	103	95
Belfast Centre	–	–	69	54	57	78	72

★ 1996 data are provisional

**Table II.8.4:
Continued**

Site	1990	1991	1992	1993	1994	1995	1996★
Birmingham Centre	–	–	86	77	82	94	90
Newcastle Centre	–	–	53	57	–	–	83
Leeds Centre	–	–	–	76	71	81	84
Bristol Centre	–	–	–	89	74	93	79
Liverpool Centre	–	–	–	50	61	81	77
Birmingham East	–	–	–	–	97	96	98
Hull Centre	–	–	–	–	98	78	81
Leicester Centre	–	–	–	–	91	115	94
S'hampton Centre	–	–	–	–	91	93	70
London Bexley	–	–	–	–	107	112	93
Swansea	–	–	–	–	–	114	87
Middlesbrough	–	–	–	–	–	96	99
London Brent	–	–	–	–	–	–	101
London Haringey	–	–	–	–	–	–	103
London Kensington and Chelsea	–	–	–	–	–	–	97
London Eltham	–	–	–	–	–	–	119
Manchester Piccadilly	–	–	–	–	–	–	82
Sheffield Centre	–	–	–	–	–	–	97
London Sutton	–	–	–	–	–	–	103
London Wandsworth	–	–	–	–	–	–	79
Wolverhampton Centre	–	–	–	–	–	–	98

★ 1996 data are provisional

Table II.8.5:
Summary of the photochemical trajectory model results for situations in which hydrocarbon and NO$_x$ emissions within the UK are eliminated, within the rest of Europe and within the whole of Europe.

Index	Base case	No precursor[e] emissions in the UK	No precursor emissions in the rest of Europe	No precursor emissions in the whole of Europe
Number of arrival points[a] >120 ppb[b]	5	2	0	0
Number of arrival points >90 ppb[c]	17	10	5	0
Number of arrival points >60 ppb[d]	41	38	21	0
Maximum calculated ozone peak, ppb	148	123	109	47
Mean peak ozone, ppb	79	71	59	46

Notes:

a. There are 50 arrival points altogether, located at the centres of the 100 km x 100 km Ordnance Survey grid squares.

b. 120 ppb represents the ozone level set as the US EPA NAAQS for a one hour mean concentration.

c. 90 ppb represents the hourly mean ozone concentration level set to trigger the provision of information to the general public in the EC Ozone Directive.

d. 60 ppb represents the hourly mean ozone concentration which is close to the running 8-hour mean ozone concentration of 50 ppb.

e. The term precursor refers to the hydrocarbon and NO$_x$ emissions from human activities.

Table II.8.6: Estimated emission reductions by 2010 compared with present day emissions – current policies

	% Reduction	
	UK	**Europe**
VOCs	40	30
NO_x	49	29
CO	56	56
SO_2	71	40

Table II.8.7: Summary of the main results from the photochemical trajectory model (ozone concentrations in ppb)

Index	Base case	25%	50%	75%	90%
Maximum	148.26	134.75	117.51	100.09	81.62
Most exposed	94.45	91.76	83.60	69.29	55.51
Pop. weighted mean	74.49	71.34	67.09	59.97	50.53
Area weighted mean	71.65	69.18	65.29	58.19	50.13

**Figure II.8.1:
Ground-level ozone
concentrations**

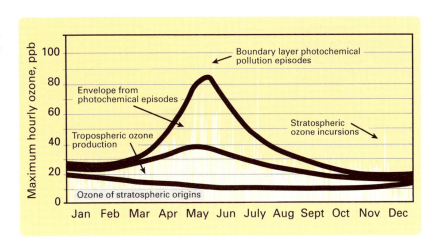

**Figure II.8.2:
Forecast ozone
concentrations (ppb)
and associated back
trajectories**

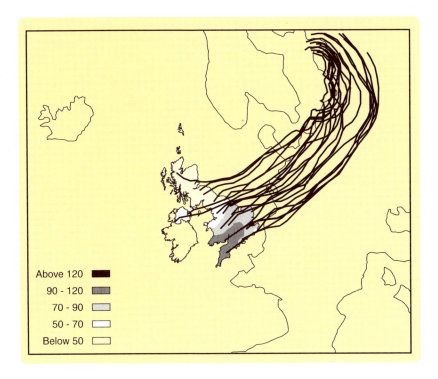

**Figure II.8.3:
Spatial distribution of
ozone hours above 60
ppb for the period April
to September 1987-1990**

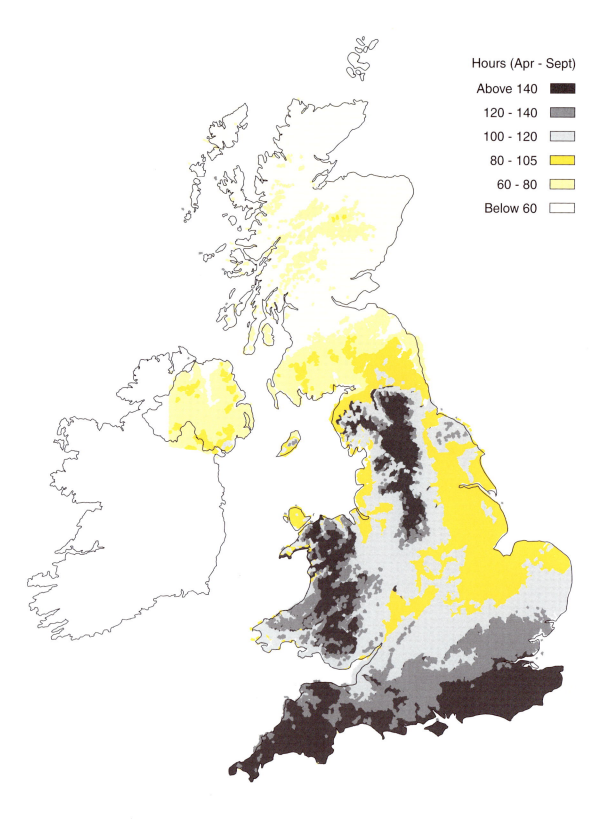

Hours (Apr - Sept)

Above 140

120 - 140

100 - 120

80 - 105

60 - 80

Below 60

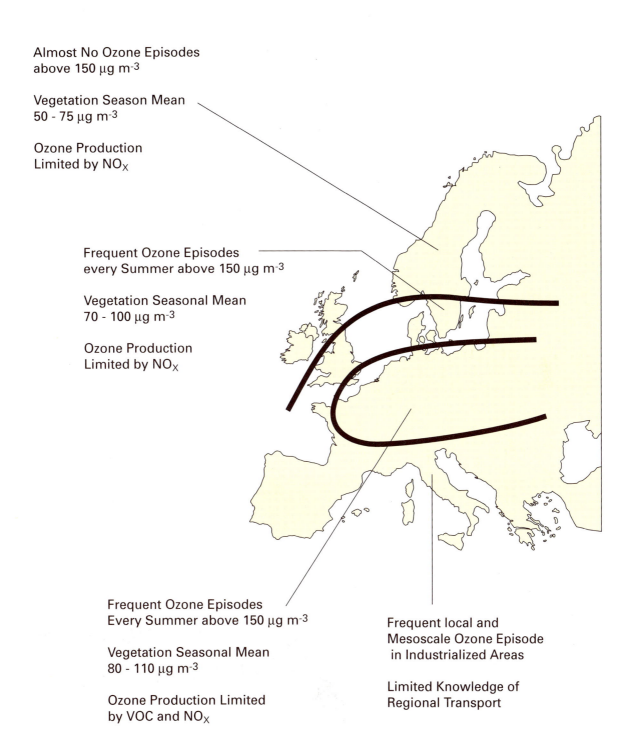

Figure II.8.4:
Photochemical oxidants
and the different air
pollution regions in
Europe

Almost No Ozone Episodes
above 150 μg m^{-3}

Vegetation Season Mean
50 - 75 μg m^{-3}

Ozone Production
Limited by NO$_X$

Frequent Ozone Episodes
every Summer above 150 μg m^{-3}

Vegetation Seasonal Mean
70 - 100 μg m^{-3}

Ozone Production
Limited by NO$_X$

Frequent Ozone Episodes
Every Summer above 150 μg m^{-3}

Vegetation Seasonal Mean
80 - 110 μg m^{-3}

Ozone Production Limited
by VOC and NO$_X$

Frequent local and
Mesoscale Ozone Episode
in Industrialized Areas

Limited Knowledge of
Regional Transport

Figure II.8.5:
Ozone formation is
'Hydrocarbons' (HC) or
'NOₓ' limited with
respect to controls in
UK and European
emmissions

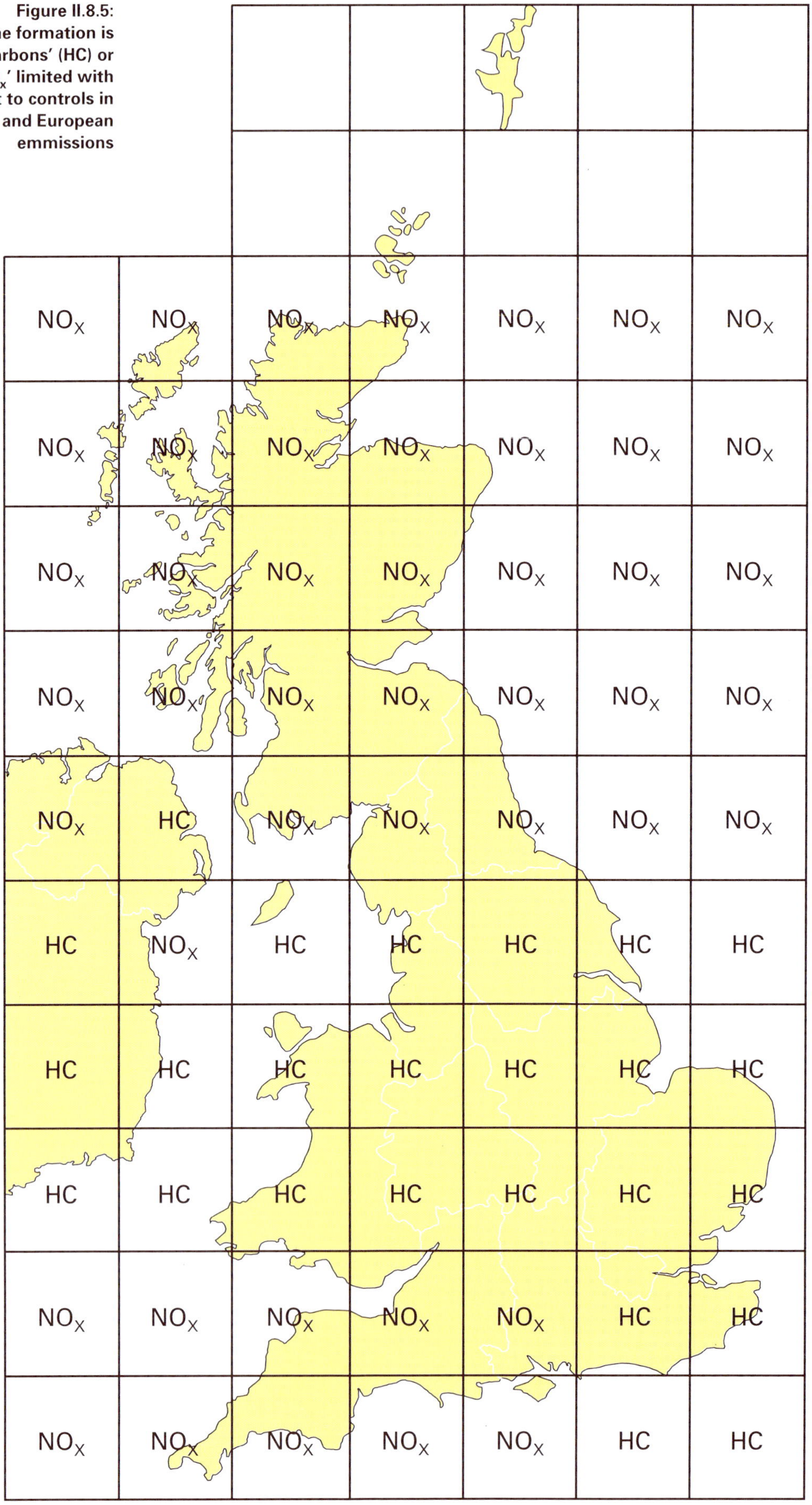

NO_X	NO_X	NO_X	NO_X	NO_X	NO_X	NO_X
NO_X	NO_X	NO_X	NO_X	NO_X	NO_X	NO_X
NO_X	NO_X	NO_X	NO_X	NO_X	NO_X	NO_X
NO_X	NO_X	NO_X	NO_X	NO_X	NO_X	NO_X
NO_X	HC	NO_X	NO_X	NO_X	NO_X	NO_X
HC	NO_X	HC	HC	HC	HC	HC
HC	HC	HC	HC	HC	HC	HC
HC	HC	HC	HC	HC	HC	HC
NO_X	NO_X	NO_X	NO_X	NO_X	HC	HC
NO_X	NO_X	NO_X	NO_X	NO_X	HC	HC

Figure II.8.6:
Assessment of control
strategies for regional
scale ozone formation

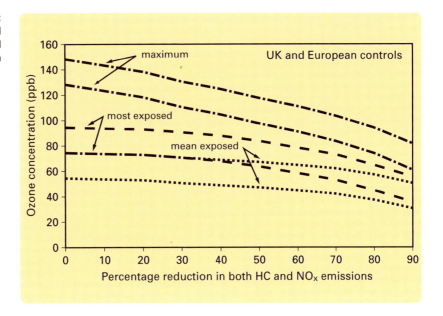

Chapter II.9: Particles

Introduction and
Sources

1. Unlike the individual gaseous pollutants discussed elsewhere in this document, which are single, well-defined substances, particulate matter in the atmosphere is composed of a wide range of materials arising from a variety of sources. Examples of man-made sources are: carbon particles from incomplete combustion; ash; recondensed metallic vapours; and so-called secondary particles, or aerosols, formed by chemical reactions in the atmosphere. As well as being emitted directly from combustion sources, man-made particles can arise from: mining; quarrying and construction operations; brake and tyre wear in motor vehicles; and from road dust resuspended by moving traffic or strong winds. Natural sources of particles include wind-blown dust and sea-salt, and biological particles such as pollens and fungal spores.

2. Average levels of particles in the air in UK towns and cities derived from domestic coal burning have decreased at some locations by as much as tenfold in the past 25-30 years following the Clean Air Act of 1956 and subsequent restrictions of coal burning in the domestic sector. Over this period, measurements of particles in the air have been made using the "Black Smoke" method whereby air is drawn through a filter paper for 24 hours and the blackness of the stain produced is then measured. This is a relatively crude and simple technique which is now being superseded by direct weighing (gravimetric) techniques which give a more direct measurement of particle concentrations in the atmosphere.

3. Although many of the obvious effects of air pollution disappeared with the earlier smogs, research over the last few years has suggested that, even at the much lower levels now found in the UK, particulate air pollution appears to be associated with a range of measures of ill-health including effects on: the respiratory and cardiovascular systems; asthma; and mortality. In the light of this evidence the Government invited the Committee on the Medical Effects of Air Pollutants (COMEAP) to advise on the health effects of particles[1] and the Expert Panel on Air Quality Standards (EPAQS) to recommend an air quality standard for particles[2].

4. The Committee and the Panel published their reports in November 1995. At the same time the Government published a preliminary response to the reports. In this, the Government adopted the level recommended by EPAQS as an provisional target for policy and gave commitments in that response to consider whether the level recommended by EPAQS should be adopted as a firm target for policy.

5. The variety of sources and forms of natural and man-made particles has been discussed above, and reference was made briefly to methods of measurement of airborne particles. Despite the usefulness of the Black Smoke method, in more recent years gravimetric methods have been developed. These methods determine the mass present of the size fraction (a part of the total amount of particles of all sizes) which is considered most likely to be deposited in the lung. In general, particles smaller than about 10 μm (1 μm = 1 micron = 1 millionth of a metre) have the greatest likelihood of reaching the lung. The gravimetric method most commonly used for measuring particles uses a size-selective inlet which collects small particles preferentially, collecting 50% of 10 μm aerodynamic diameter particles,

---

[1] Committee on the Medical Effects of Air Pollutants, *Non-biological Particles and Health*, HMSO, 1995

[2] Expert Panel on Air Quality Standards, Particles, HMSO, London, 1995.

more than 95% of 5 μm particles and less than 5% of 20 μm particles. The resultant mass of material is known as PM_{10}. This material is thus defined solely by physical characteristics rather than chemical composition, and, moreover, both of these attributes can vary from place to place, unlike a gaseous pollutant such as nitrogen dioxide. Broadly speaking, however, man-made particles tend to lie in the smaller size ranges, below about 10 μm. Naturally-generated particles tend to be larger, but will nonetheless be captured to some extent by PM_{10} samplers.

6. Because of the wide variety of sources, sizes and atmospheric behaviour of particles, accounting for the sources and reconciling them with measured air concentrations is more difficult than with a more straightforward pollutant like carbon monoxide. Nonetheless, much progress has been made in the past year or so and is summarised in the third report of the Quality of Urban Air Review Group (QUARG)[3].

7. Estimates of primary mass emissions of PM_{10} from man-made sources are shown in Table II.9.1 for the UK as a whole. The largest single source is road transport which accounts for 25% of the total. However, large-scale inventories for the UK do not provide an accurate description of the relative contributions to PM_{10} levels in town and cities, for several reasons. The emissions from power stations mostly arise from rural locations and much of the emissions from industrial sources, mining and quarrying comes from

Table II.9.1
Emissions of PM_{10} in the UK, 1995

Source	Emissions (kilotonnes)*	Percentage of Total**
Power Stations (Fossil Fuelled)	34	15
Domestic	20	9
Commercial/public service	5	2
Refineries	7	3
Iron and Steel	20	9
Other Industrial Combustion	15	6
Construction	4	2
Industrial Processes	30	13
Mining & Quarrying	29	12
Extraction and distribution of Fossil Fuels	0	0
Solvent Use	0	0
Road Transport:		
Diesel	43	18
Petrol	11	5
Non–exhaust (Tyres and Brakes)	5	2
Other Transport	7	3
Waste Treatment and Disposal	0	0
Agriculture	2	<1
Total	**232**	**100**

*   Figures rounded to nearest kilotonne
**   Figures rounded to nearest 1%

[3] Quality of Urban Air Review Group, *Airborne Particulate Matter in the United Kingdom*, London, 1996.

sites outside towns. These sources will, of course, contribute to urban levels, but generally in smaller proportions than those given in Table II.9.1. Similarly, the relatively large proportion of total UK emissions arising from the domestic sector occurs mainly from coal burning and only makes a significant contribution in a relatively small number of locations. However, as is clear from the measurement in Belfast, widespread domestic solid fuel use can lead to high PM_{10} levels. Moreover, natural sources of PM_{10} are not included in the inventory in Table II.9.1 and these will clearly contribute to measured urban levels. Of the man-made sources, however, road transport will be the dominant contributor to PM_{10} levels in the air in most major towns and cities in the UK.

8. An analysis by QUARG concludes that, in winter, road transport typically contributes 40-50% of urban PM_{10} levels. QUARG further concludes that the percentage contribution is typically higher in winter episodes and presents analyses which demonstrate that when daily PM_{10} levels are above 50 $\mu g/m^3$, the contribution from traffic is the range of 75%-85% of total PM_{10}.

9. Levels of PM_{10} are generally highest in winter, but levels above 50 $\mu g/m^3$ can also occur in summer. In the still conditions responsible for 'summer smog', vehicle emissions will make a contribution to overall PM_{10} levels, but an important component is that arising from so-called secondary particles, or aerosols. These are particles, typically composed of ammonium sulphate and nitrate which are formed photochemically from emissions of sulphur and oxides of nitrogen emissions from industry, power generation and transport sources across Europe, including the UK, and which can travel long distances in the atmosphere. The science is not yet available to quantify the relative contributions from secondary aerosol and road traffic. Although the contribution from the latter is likely to be lower than in winter episodes, it is still likely to be substantial.

Health Effects:
Standards and
Guidelines

10. In the 1950s and 1960s, the effects of the high levels of airborne particles, in combination with sulphur dioxide, in the notorious smogs were clear. In the last few years research has demonstrated associations between a range of health outcomes - respiratory and cardiovascular effects and mortality - and various measures of particle concentrations at much lower levels than in the earlier smog studies. Initially, most of this work appeared in the USA, but recently similar findings have been reported in Europe and in South America.

11. As has been noted above, this work has recently been assessed by COMEAP advising the Department of Health, who considered that it would be imprudent not to regard the reported associations as causal. EPAQS, using this assessment as a basis, and other more recent work carried out in Birmingham, addressed the problem of setting a standard when the published work was unable to identify a no-effect threshold. The Panel adopted a risk management approach, similar in concept to that which it used for genotoxic carcinogens, and sought to identify a level at which the effects on the population as a whole would be relatively small. Using this approach, EPAQS identified a level at which one might expect one additional hospital admission (for respiratory disorders) per day in a population of one million. This value, 50 $\mu g/m^3$ as a running 24-hour mean, was recommended as a standard for PM_{10}. EPAQS further recommended that annual average PM_{10} concentrations should be reduced.

12. In their review of the health effects of non-biological particles, COMEAP noted that "in the absence of strong evidence on the relative effects of

different particles within the respirable range, it seems reasonable, at present, to base policy on PM_{10} measurements". They also noted that "it has been suggested, but by no means proven that ultrafine particles (< 0.05 μm diameter) may play a role". EPAQS accordingly made their recommendation for particles based on the PM_{10} fraction, but noted that the smaller $PM_{2.5}$ fraction "may eventually prove to be of greater health significance". In view of this, and since the Government has recently commissioned further research in this area, including studying the possible mechanisms of action of fine particles and investigating concentrations of fine particles and particle numbers across the UK, EPAQS recommended that the air quality standard for PM_{10} should be reviewed within the next five years.

Current Air Quality

13. In 1992 the Department of the Environment established a network of automatic PM_{10} analysers, reporting concentrations from city centre locations every hour. In January 1997 there were 35 sites monitoring PM_{10} in this way. Figure II.9.1 shows the maximum daily average concentrations recorded for each month and site between 1992 and 1994. It can be seen that concentrations are highest in the winter months and lowest in the summer. However, the difference is less than that seen for other motor vehicle-derived pollutants such as carbon monoxide or oxides of nitrogen. This is because there is another important source of particles during summer, the secondary particles described earlier. Table II.9.2 shows the number of days from 1992 to 1995 when the running 24-hour average PM_{10} concentrations exceeded 50 μg/m³.

Table II.9.2:
Number of days when
the PM_{10} concentration
of 50 μg/m³ as a running
24-hour mean was
exceeded.

Site	1992	No of days exceeding 1993	1994	1995
London Bloomsbury	43	57	39	46
Edinburgh	6	4	3	18
Cardiff	17	40	91†	24
Belfast	40	80	32	40
Birmingham	37	39	23	24
Newcastle	38	32	39	20
Leeds	–	40	44	38
Bristol	–	50	30	27
Liverpool	–	28	34	52
Birmingham East	–	–	19	20
Hull	–	–	30	21
Leicester	–	–	17	8
Southampton	–	–	16	13
Bexley	–	–	18	33
Swansea	–	–	–	41
Middlesbrough	–	–	–	28

† Temporary construction activity near the measurement site

In 1995 the greatest number of exceedences occurred in London, Belfast (where domestic solid fuel use is widespread), Swansea, and Liverpool, with the smallest in Leicester. Current levels of PM_{10} in the UK are such that the standard is exceeded on typically 10% or so of days in a year. The number of days exceeding the standard during this period has ranged from 3 to 91, the latter number at the Cardiff site where the clear additional effect of particle emissions from a period of local construction activity (which took place in 1994 and has now ceased) adding to the road traffic emissions can be seen.

14. The highest daily means are observed at Belfast where domestic solid fuel use is widespread. There will also be other areas of the UK where this is the case and where traffic is not the single dominant source. In addition to construction activity, industrial sources and mineral working, whilst not significant contributors to peak PM_{10} levels during episodes across the country as a whole, are among possible causes of a localised episode at any time of the year.

The Strategy

15. There are already reduction measures in place for most of the man-made sources of PM_{10} described above. Over the next 10 years they should substantially reduce particle emissions in the UK from these sources. This should in turn lead to much lower concentrations of PM_{10} in the air. Most of the reduction will come from primary transport emissions and secondary aerosol.

Transport

16. The most important measures are:

- 'EURO I' vehicle emissions Directives which set limits for the emissions of nitrogen oxides, hydrocarbons, carbon monoxide and particles (diesel vehicles only). In practice EURO I required the fitting of catalytic converters to all new petrol cars;

- Directive 93/441/EEC, passenger cars (from 31/12/1992);

- Directive 93/59/EC, light vans (from 1/10/1994);

- Directive 91/542/EEC, heavy duty vehicles (from 1/10/1993);

- 'EURO II' vehicle emissions Directives which tighten further the limits in EURO I;

- Directive 94/12/EC, passenger cars (from 1/1/1997);

- Directive 96/69/EC, light vans (from 1/10/97);

- Directive 91/542/EEC, heavy duty vehicles (from 1/10/1996);

- reduction of the EC sulphur level in diesel fuel from 0.3 to 0.2% in 1994 and from 0.2% to 0.05% in 1996 (around 2.5% reduction in particle emissions for cars and vans and 13% for lorries and buses);

- tightening of UK diesel-in-use smoke standards (1995);

- Budget measures in 1995 and 1996 were aimed primarily at reducing emissions from the transport sector and will contribute to reductions in particle emissions. These measures are discussed in more detail in Chapter 6 of Part I.

Secondary aerosols

17. The most important measures here are:

- ongoing implementation of the UNECE Convention on Long Range Transboundary Air Pollution Second Sulphur Protocol. This will lead to a 70% reduction in UK emissions of sulphur dioxide by 2005 80% by 2010 relative to 1980 baseline;

■ further reductions of NO_x (and possibly sulphur) emissions in the UNECE Second NO_x Protocol.

Table II.9.3: 99th Percentile of the maximum daily 24-hour running means of PM_{10} 1992-1995 ($\mu g/m^3$)

Site	1992	1993	1994	1995
London Bloomsbury	85	81	80	72
Edinburgh	72	52	49	67
Cardiff	69	85	92	59
Belfast	214	110	162	181
Birmingham Centre	85	80	74	78
Newcastle	74	86	70	72
Leeds	-	78	113	79
Bristol	-	77	72	62
Liverpool	-	100	84	70
Birmingham East	-	35	64	85
Hull	-	-	76	68
Leicester	-	-	61	52
Southampton	-	-	63	64
Bexley	-	-	70	76
Swansea	-	-	46	70
Middlesbrough	-	-	-	71

- monitoring not started

Other sources

18. Other measures to help reduce particle emissions include:

■ ongoing implementation of the EC Large Combustion Plants Directive (also helping reduce secondary aerosols);

■ the implications of the Environmental Protection Act 1990 - IPC, Local Air Pollution Control System (LAPC) (again helping reduce secondary aerosols);

■ smoke control areas (effect in reducing particle emissions mainly complete, but some programmes are still to be finished).

Objectives

19. Notwithstanding these measures, there will inevitably be occasions when the standard will be exceeded for reasons which would either require disproportionately expensive measures, or would arise for social and cultural reasons. The Government therefore has decided to adopt a percentile approach to the management of PM_{10} levels, and to adopt as a provisional objective 50 $\mu g/m^3$ running 24-hour mean as a 99th percentile, (i.e. on all but four days per year, assuming perfect operation of the

monitoring station with 100% data capture in the year). Here, the approach in calculating percentiles for a running mean is to assign the highest running 24-hour mean in any one day to that day and calculate the percentiles of these 365 values in a year.

20. Data on the 99th percentiles for PM_{10} since measurements began are shown in Table II.9.3. From this data it can be seen that, leaving aside Belfast (which is unique in that domestic solid fuel makes a major contribution to PM_{10} levels), the highest values of the 99th percentile are about 100-113 $\mu g/m^3$.

21. The extent of the emission reductions required to achieve the objective for PM_{10} throughout urban areas can be calculated using the analysis provided by QUARG. It is possible to estimate how much abatement of the major sources of PM_{10} is required during winter smog periods in order to reduce a concentration of 100 $\mu g/m^3$, for example, to one of 50 $\mu g/m^3$. Measurements suggest that the coarse (i.e. larger) fraction of PM_{10} shows little increase during high pollution periods and is steady at about 15-20 $\mu g/m^3$. The concentration of secondary PM_{10} is approximately 10 $\mu g/m^3$. Thus, allowing for the small fraction of the coarse particles associated with vehicle emissions and secondary PM_{10}, there is a background of about 25 $\mu g/m^3$ which is not directly associated with vehicle emissions and will not respond to control of this source. Therefore, to reduce a winter smog concentration of 100 $\mu g/m^3$ to below 50 $\mu g/m^3$ will require a reduction in the vehicle emitted component from 75 $\mu g/m^3$ to 25 $\mu g/m^3$ - a reduction of 67%. However, reductions in the precursors of secondary PM_{10} agreed in the UNECE Sulphur Protocol have been estimated to deliver a reduction in the secondary component of about 4 $\mu g/m^3$ by the year 2010. Taking this into account, the extent of abatement necessary for vehicle emissions is then 61-67% on 1995 emissions.

22. Figure II.9.2 shows projected emissions of PM_{10} from urban road traffic to 2025 on the basis of current policies. The projected emissions are, however, subject to some uncertainty, and are likely to be underestimates of emissions in future years. This is because no quantitative information is available on the degradation of emission performance in service throughout a vehicle's life. A small amount of data is available for gaseous pollutants, and this has been used in Chapter II.7 on nitrogen dioxide, but no data are available on particles. Accordingly, since it would be quite unrealistic to assume no degradation in performance, an assumption has been made that emissions from diesel cars and light vans degrade linearly in service, such that after 80,000 km their emissions are a factor of two greater than the regulated value for the new vehicle. For post-2000 vehicles a similar approach is adopted, but using a factor of 1.2. No degradation in performance (in terms of particle emissions) has been assumed for petrol cars.

23. Furthermore, the projections assume that there is no increase in the penetration of new diesel cars into new car sales beyond the current value of 20%. The projections use the mean of the National Road Traffic Forecast high and low forecasts.

24. The dramatic reduction in emissions in these projections, amounting to a decrease of over 40% by 2005 compared with 1995, should lead to significantly lower levels of particles in the air. This will be reflected particularly in wintertime pollution episodes.

25. Current estimates are that the anticipated decline in emissions should go far towards reducing exceedences of the specific objective. After about 2005 emissions should begin to increase again as forecast increases in traffic activity outweigh the benefits obtained by improved technology. This would run counter to the principles of sustainable development, particularly for a pollutant for which no zero-effect threshold can currently be identified. However, further improvements in vehicle technology and fuels should by then be capable of delivering even greater reductions than those projected in Figure II.9.2.

26. Formal proposals have been adopted by the European Commission on the next phase of vehicle and fuel measures to reduce emissions of particles and gaseous pollutants in the so-called Stage III limits. The UK has argued for substantial emission reductions, consistent both with attaining the level recommended by EPAQS on a percentile basis (see below) in wintertime episodes in the UK by 2005, and maintaining at least this level of air quality (in terms of particles) thereafter. The European Auto-Oil Programme estimated some annual costs associated with achieving reductions in oxides of nitrogen and particle emissions. Not only will these measures reduce particle emissions, they will also substantially reduce emissions of VOCs, benzene, 1,3-butadiene, carbon monoxide, and nitrogen oxides.

27. Other measures which have been costed include conversion of diesel engines to operate with alternative fuels such as Compressed Natural Gas (CNG) and Liquefied Petroleum Gas (LPG). This can result in reductions in the emissions of particles and NO_x by up to 70% per vehicle.

28. Abatement costs (see Annexes 1 and 2) can be set against estimates of adverse effects from particles. Department of Health assessments suggest that particles are responsible for several thousand advanced deaths each year, for substantial numbers - of the order of 10-20,000 - of hospital admissions, and for many thousands of instances of illness and reduced activity. Estimates of the value of reducing irritation from the dirt and odour associated with particle emissions are necessarily subjective. However, there will also be considerable reductions in the amount of money spent in cleaning soiled buildings.

Conclusions 29. The Government accepts the advice of COMEAP that it would be imprudent not to regard the associations between mortality and morbidity and particle concentrations as causal. It also accepts the recommendations of EPAQS and will adopt the value of 50 $\mu g/m^3$ running 24-hour mean as the standard for PM_{10}. The Government will also adopt this value as a provisional objective to be achieved at the 99th percentile (i.e. on all but 4 days per year assuming perfect operation of the monitoring station with 100% data capture in the year by 2005). It further intends to ensure that annual average levels of PM_{10} are progressively reduced.

30. In considering where the objective for PM_{10} should apply, it is clear that over a 24-hour period, patterns of activity will be such that a range of micro environments will be experienced, potentially including heavily trafficked streets at one extreme, and cleaner rural areas as well as indoor locations which may or may not include indoor sources of particles. As was noted in the Introduction to Part II, the Government intends to undertake further research on personal exposures, and in advance of this work, it is prudent to assume that roadside environments are likely to provide an upper limit to exposures over 24 hours.

31. The Government, therefore, has decided that the objective for PM_{10} should apply in the following non-occupational, near-ground level outdoor locations: background locations; roadside locations; and other areas of elevated PM_{10} concentrations where a person might reasonably be expected to be exposed (e.g. in the vicinity of housing, schools or hospitals etc) over a 24 hour period. This approach will be reviewed in the light of further research into patterns of personal exposure. Consideration of potential exceedences of the provisional PM_{10} objective at potentially high concentration locations should be carried out in conjunction with data from urban or other appropriate background locations.

Figure II.9.1:
Maximum daily average
PM_{10} concentrations at
UK sites 1992-1994

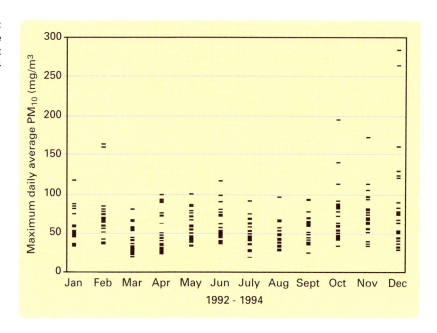

Figure II.9.2:
Future UK urban road
transport emissions of
PM_{10} (current policies;
diesel car sales at 20%
of all new car sales)

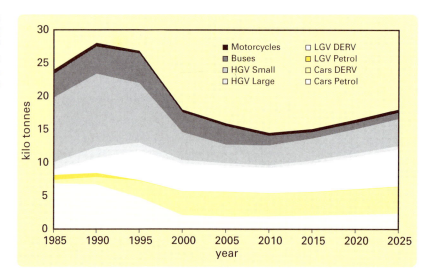

Chapter II.10: Sulphur Dioxide

Introduction and Sources

1. Sulphur dioxide is a gas at normal temperature and pressure. It dissolves in water to give an acidic solution which is readily oxidised to sulphuric acid. In the United Kingdom the predominant source of sulphur dioxide is from the combustion of sulphur-containing fossil fuels, principally coal and heavy oils. It is an irritant when it is inhaled, because of its acidic nature, and high concentrations may cause breathing difficulties in people exposed to it. Recent studies have shown that people suffering from asthma may be especially susceptible to the adverse effects of sulphur dioxide and that, within the range of concentrations that occur in pollution episodes, it may provoke attacks of asthma. Sulphur dioxide can also lead to direct adverse effects in vegetation, with trees, crops, and other native flora all known to be affected by the kind of levels that are sometimes found in the UK.

2. For the first half of this century, emissions of sulphur dioxide were dominated by the combustion of coal, not only in the domestic sector, but also in commercial and industrial premises, and in power stations which were situated predominantly within towns and cities. Following the smogs in the 1950s and the Clean Air Act of 1956, this pattern changed as cleaner fuels replaced coal in the domestic, commercial and industrial sectors, and power generation was concentrated in much larger, more efficient stations situated in rural areas.

Figure II.10.1: Mean annual concentrations of sulphur dioxide

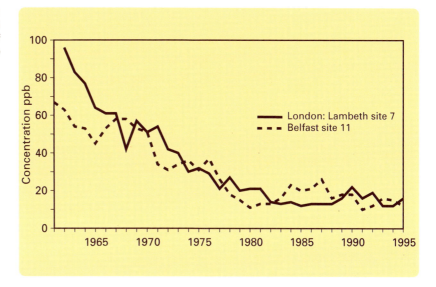

3. As a result, national emissions of sulphur dioxide have decreased by 63% since 1970, and by 52% since 1980. Urban smogs like those in the 1950s are now a thing of the past, and average levels of sulphur dioxide have decreased by around fivefold since the 1960s (see Figure II.10.1). Emissions are now dominated by fossil-fuelled power stations which currently account for 67% of the national total. Table II.10.1 shows the national emission inventory for 1995. Although the use of solid fuel outside the power generation sector is small overall, there are still some areas of the UK, most notably in Belfast and some areas associated with coal mining, where domestic coal and smokeless solid fuel is burned in significant quantities.

4. Overall therefore, emissions of sulphur dioxide are dominated by a relatively small number of large emitters (in 1995, for example, 44% of all sulphur dioxide emissions arose from just nine power stations). There is also a significant emission from the industrial sector, including refineries.

Chapter II.10: Sulphur Dioxide

Table II.10.1: Emissions of Sulphur Dioxide in the UK, 1995

Source Category	Emissions* (kilotonnes)		Percent contribution**	
1. Public Power etc.	1588		67	
Coal		1443		61
Fuel oil		145		6
2. Commercial, institutional, residential combustion plants	123		5	
Domestic		68		3
Other		55		2
3. Industrial combustion plants	437		19	
Refineries		123		5
Iron and steel		62		3
Other		252		11
4. Road transport	51		2	
Petrol exhaust		16		1
Diesel		35		1
5. Other transport	69		3	
6. Other	97		4	
TOTAL		**2365**		**100**

*   Figures rounded to nearest kilotonne

**   Figures rounded to nearest %

In contrast to other pollutants, transport emissions of sulphur dioxide are relatively unimportant nationally, but the combustion of diesel fuel can make a significant contribution to background levels in urban areas.

Health Effects: Standards and Guidelines

5. Until relatively recently, air quality guidelines and standards for sulphur dioxide were expressed in conjunction with values for Black Smoke or other measures of particles. This originated from the pioneering work in the UK in the smogs of the 1950s and 1960s, and forms the basis of the current Directive (80/779/EEC) on smoke and sulphur dioxide, and also most of the WHO guidelines published in 1987. More recent research has addressed the effects of sulphur dioxide acting alone, and exposures of the order of minutes have been shown to exhibit adverse effects on human health. Such research has been incorporated into two influential health based reviews[1,2]:

■ The Expert Panel on Air Quality Standards (EPAQS) have recommended an air quality standard for sulphur dioxide of 100 ppb, measured over a 15 minute averaging period.

[1]Expert Panel on Air Quality Standards, *Sulphur Dioxide*, HMSO, London. 1995.

[2]WHO Regional Office for Europe, *Update and Revision of the Air Quality Guidelines for Europe - Meeting of the Working Group "Classical" Air Pollutants, Bilthoven, The Netherlands, 11 - 14 October 1994*, WHO, Bilthoven, 1994

■ The World Health Organisation (WHO) guidelines set a figure for a short averaging period of 175 ppb (500 μg/m^3)for a 10-minute mean concentration. A recent review of these guidelines recommended the retention of this figure, but the removal of a previous guideline based on an hourly mean.

6. WHO have also retained their previous 24 hour guideline of 44 ppb (125 μg/m^3) and annual average guideline of 17 ppb (50 μg/m^3). These figures were derived from epidemiological studies.

7. These recommendations are based purely on health considerations - they aim to set levels below which it is unlikely that human health effects will be seen, even in relatively sensitive subjects, i.e. brittle asthmatics in the case of sulphur dioxide.

How the Guidelines were Derived

7. The WHO 10 minute guideline was derived by accepting that the lowest-observed- effect level for effects of concern to health reported from chamber studies was 350 ppb. Because the range of people likely to take part in such studies is limited, an uncertainty factor of 2 was applied to take into account sensitive individuals, including those suffering from asthma, and a guideline of 175 ppb obtained. EPAQS adopted a different approach in that they identified from a review of the data 200ppb as a concentration below which effects of clinical significance were unlikely in sensitive individuals, eg those suffering from asthma.

8. EPAQS considered carefully the averaging time appropriate for the standard. It was noted that the effects of sulphur dioxide appeared almost immediately at the start of exposure. Whilst it was agreed that an averaging time of just a few minutes might be desirable, it was agreed that a 15 minute averaging time represented an acceptable compromise between desirability and practicability.

9. EPAQS felt that the figure of 200 ppb would be appropriate for very short averaging times: perhaps a few minutes. It was also agreed that a 15 minute averaging period could conceal short lived peak concentrations and thus the figure recommended as a 15 minute average should be lower than that seen as acceptable for a shorter averaging period, ie should be less than 200 ppb.

10. Examination of data, collected in areas where short lived high concentrations of sulphur dioxide are likely, showed that the peak concentration could exceed that 15 minute average by a factor of 2. This factor was applied to the figure of 200 ppb and a standard of 100 ppb, 15 minute averaging time, produced.

11. The WHO 24 hour and annual average guidelines are both derived from epidemiological studies in which people had been exposed to a mixture of pollutants, and, in particular, sulphur dioxide/particulate matter mixtures typical of the pollutant mix arising from extensive domestic solid fuel use. EPAQS were less convinced of the need for longer term standards and only recommended the short term standard above. It is worth noting, however, that there is concern that the latest epidemiology studies (APHEA - a European wide study in 12 cities including London) could lead to the conclusion that significantly increased mortality can be associated with sulphur dioxide concentrations lower than the current WHO 24 hour guidelines. This new

[3]*Reducing National Emissions of Sulphur Dioxide: A Strategy for the United Kingdom,* DoE, 1996.

**Figure II.10.2:
Estimated annual mean
sulphur dioxide
concentration,
1994 (µg/m³)**

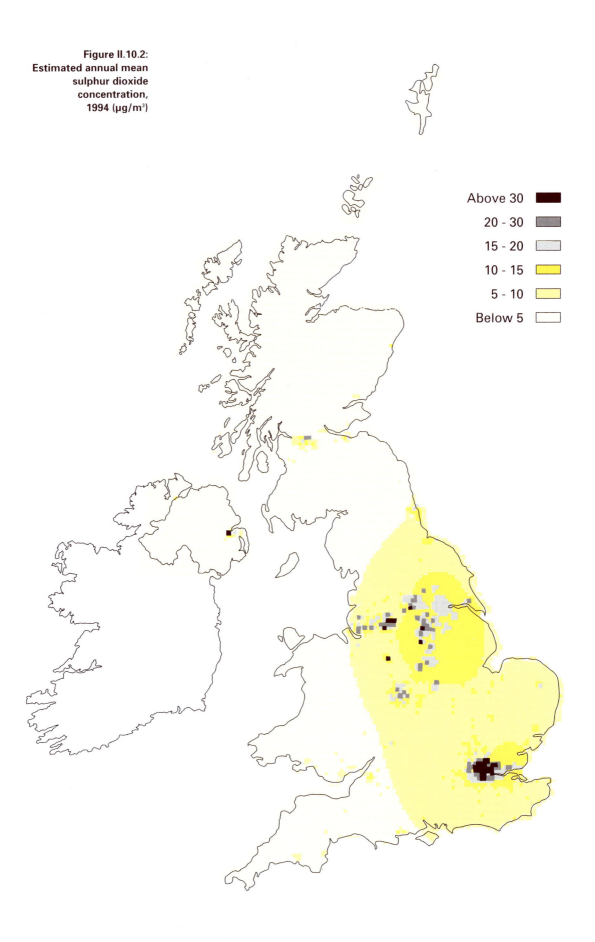

Above 30
20 - 30
15 - 20
10 - 15
5 - 10
Below 5

evidence will be considered at the first review point in 1999. It is not known what effect, if any, this would have on the EPAQS judgement that the evidence for longer term health effects was weak, and that therefore a longer term standard averaging was not appropriate. However it does add further weight to the concern over the health effects of sulphur dioxide.

12. In conclusion, the experimental evidence points to sensitive individuals being affected by peak concentrations of sulphur dioxide at levels around 200 ppb. EPAQS have correctly translated this short term peak into a 15 minute standard of 100 ppb. By contrast, the WHO recommendation effectively ignores the possibility of peaks within their 10 minute averaging time.

Current Air Quality

13. There are four major national networks currently measuring sulphur dioxide concentrations: the Basic Urban Network (BUN), the EC Directive Network, the Rural Network, and the Automatic Urban Network (AUN). Of these, only the AUN measures sulphur dioxide concentrations at the time resolution required for comparison with short term health based objectives. Nevertheless, the other networks provide useful background data for identifying areas of high and low sulphur dioxide concentrations.

14. The Basic Urban Network (BUN) and the EC Directive Network have developed from the National Smoke and Sulphur Dioxide Survey - a large network for the measurement of sulphur dioxide concentrations which was established in the early 1960s. The measurements are made on a 24-hour average basis using a standard peroxide bubbler method. In 1995/96 there were 151 and 159 sites operational in the BUN and EC Directive Network respectively. Some sites are in both networks and a few sites are operated in other networks - in total 225 sites were operational in 1995/96. The Rural Sulphur Dioxide Network was established in 1991 and measures concentrations at 38 sites across the UK on a weekly basis using a standard peroxide bubbler method.

15. Each of these networks has its own rationale. The BUN was established to provide long term trend information on sulphur dioxide (and black smoke) levels across a range of towns and cities of different sizes and pollution levels. The EC Directive Network was established to monitor compliance with the EC Directive on smoke and sulphur dioxide. The sites chosen were in areas predicted to have the highest sulphur dioxide levels. As air quality has improved and sites have moved into compliance, the size of the network has reduced. The Rural Network was established principally to allow the quantification of dry deposition of sulphur to ecosystems across the UK, to complement the wet deposition estimates obtained from the monitoring of rainfall composition. It also provides information on background levels across the UK.

16. Between 1993/94 - 1995/96, the average concentrations over the BUN and EC Directive Network were 9.0 ppb and 10.1 ppb respectively. The lowest values (less than 8 ppb) were generally observed away from major conurbations. Higher values (8 to 17 ppb) were typical in and around the main centres of population. However, the highest values (greater than 17 ppb) occurred both within conurbations and away from them. There was no correlation between size of town and average concentrations; in many cases the highest levels were found in small communities. Annual mean concentrations of between 17 - 24 ppb were observed in areas associated with coal mining where the use of coal is still widespread. Similar

Table II.10.2: 1993 - 95 Calendar Year 15 Minute Mean Statistics for Sulphur Dioxide.

Site	Max 15 min. min. mean ppb	No. of 15minute means >100ppb			No. of 15 minute means >175 ppb		
Year		93	94	95	93	94	95
Strath Vaich (N Scotland)	24	0	0	0	0	0	0
Edinburgh (Centre)	218	50	30	19	4	2	1
Newcastle (Centre)	322	34	43	97	1	2	10
Sunderland	139	34	18	2	0	0	0
Middlesbrough	150			24			0
Belfast (Centre)	495	449	401	357	120	116	112
Belfast (East)	569	852	464	590	242	157	299
Leeds (Centre)	513	99	164	81	9	48	24
Hull (Centre)	321		31	55		4	4
Barnsley 12	212		16	71		0	13
Liverpool (Centre)	307	185	129	56	20	22	3
Ladybower (Derbyshire)	260	19	33	42	2	7	8
Leicester (Centre)	212		17	5		1	0
Birmingham (Centre)	221	46	14	23	8	0	12
Birmingham (East)	182		1	18		0	3
Stevenage	128	12	0		0	0	
Swansea	120			3			0
London (Bloomsbury)	306	121	38	63	20	2	5
London (Bridge Place)	239	67	28	42	7	2	7
London (Cromwell Rd)	259	71	27	39	10	2	11
Cardiff (Centre)	175	10	0	4	0	0	1
Bristol (Centre)	197	14	3	1	1	0	0
London Bexley	481		61	164		10	48
Lullington Heath (E Sussex)	64	0	0	0	0	0	0
Southampton (Centre)	134		2	0		0	0

concentrations have been observed in Northern Ireland, particularly in Belfast, where the use of coal and coal products is again widespread in the domestic sector.

17. The data from the BUN and the Rural Network can be combined and interpolated to give a map of annual average concentrations in the UK. The map for 1994 is presented in Figure II.10.2. The highest annual

**Figure II.10.3:
The time-of-day
dependence of the
occurrence of
exceedences of EPAQS
SO$_2$ standard during
1993 and 1994 at
Ladybower, Derbyshire**

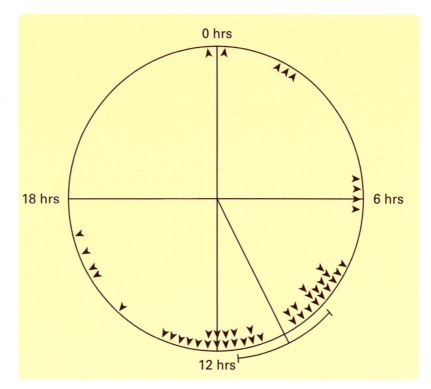

average concentrations of sulphur dioxide are seen to be in central England, which is broadly consistent with the location of the highest emissions of sulphur dioxide.

18. Continuously recording automatic analysers are necessary to measure sulphur dioxide concentrations over short periods. Measurements using such instruments have been made on a small scale since the 1970s, but since 1992 the number of sites has increased with the establishment of the Automatic Urban Network. There are currently (January 1997) 44 stations operating in the network, including six rural sites. The network is currently being expanded, and now incorporates many sites run by local authorities. By the end of 1997, it is expected that the network will consist of 65 sites, including seven in rural areas.

19. The automatic measurements have revealed that, despite the reductions in annual average concentrations, peak concentrations are still such that the level recommended by EPAQS is widely exceeded. This occurs, not just at urban sites, but also in rural areas where the site is liable to be influenced by the plume of a single large emitter of sulphur dioxide.

20. There are three main causes of these exceedences, which, while they usually interact with each other, are each capable of causing exceedences in both time and area in isolation from the other two:

21. Plumes from large point sources (combustion plants with rated thermal input greater than 50 MW) may cause relatively infrequent exceedences, but, when they do occur, can cover large areas with very high concentrations,

22. Plumes from smaller point sources (< 50 MW) may cause more frequent exceedences, but over smaller areas and with lower concentrations,

23. Diffuse low-level sources, particularly in areas where coal remains a popular domestic heating fuel, can lead to relatively high annual means,

with particularly high levels experienced under adverse weather conditions in the same manner as, although less severe than, the smogs of the past.

24. Of all the continuous sulphur dioxide monitoring sites, only the two most remote sites at Strath Vaich in the Highlands and Lullington Heath on the south coast of England, did not record any exceedences of the EPAQS recommendation between 1993 and 1995. Exceedences occurred with the highest frequency and with the highest severity in the industrial and domestic black fuel burning areas: e.g. Belfast, Liverpool, Leeds. Exceedences were recorded at all the other urban background and rural sites, although the number and severity of these exceedences vary from site to site and from year to year.

25. Direct comparison with the WHO short-term guideline is not strictly possible as the AUN does not record data on a 10 minute basis. Indeed, since there are no contemporaneous measurements on a 10 and 15 minute basis from the AUN, it is not possible to accurately define a conversion factor. However, it can be assumed that there is unlikely to be much difference between concentrations averaged over the two periods. Comparing the data for 1993 - 95 AUN 15 minute averaged data against 175 ppb shows that many sites recorded few or even no exceedences each year. As with the EPAQS comparison, the greatest exceedences, both in numbers and in magnitude, occurred in the domestic and industrial black fuel burning areas.

26. The 1993 - 95 exceedence data for the EPAQS recommendation and WHO short term guideline are summarised in Table II.10.2.

Large Combustion Sources

27. It appears that the highest peaks in sulphur dioxide concentrations observed in the United Kingdom are mainly caused by the plumes from large combustion plants reaching ground level infrequently, but during particular meteorological conditions. These large sulphur dioxide sources can therefore contribute significantly to elevated sulphur dioxide concentrations despite contributing little to long term means.

28. The 15-minute air quality data for the remote rural Ladybower site in Derbyshire for 1993 and 1994 have been studied in some detail to understand the seasonal and diurnal patterns of occurrence of the EPAQS exceedences. During this time, 49 exceedences occurred on 14 separate days. Two of the days were during winter, six were in autumn, with the remaining days in spring and early summer.

29. Figure II.10.3 pools all the exceedences of the 15-minute EPAQS recommendation for Ladybower during 1993 and 1994 and plots the time of day of the 49 exceedences. Two main groups of exceedences are found in this plot: one main group during the morning between 08.00 and 10.00 hrs, with a secondary group around midday. The morning group contains over half the exceedences and the midday group about one third. There are less frequent but still discernible groups in the mid-afternoon, around midnight, and in the early hours of the morning. Both the two main groups of exceedences gave similar peak 15-minute mean sulphur dioxide concentrations of about 255 ppb.

30. The diurnal pattern of exceedences of the EPAQS recommendation as revealed in Figure II.10.3 strongly suggests two mechanisms for their occurrence. The most frequent group occurs in the morning and covers all seasons of the year and is associated with the erosion of the nocturnal

stable layers by solar heating after sunrise. Large combustion plant plumes will generally lie above any nocturnal stable layer and fumigate the shallow boundary layer during the morning as the boundary layer rises. Peak sulphur dioxide concentrations are thus observed at ground level shortly after sunrise. Because the plumes have been above the nocturnal stable layer, they may have undergone little dispersion during a considerable travel time at night. Fumigation events may potentially occur at considerable downwind distances.

31. The second most frequent group occurs around midday and in the early afternoon and is found on warm, sunny and anticyclonic days during the Spring, Summer and early Autumn. This group is associated with convection. Convective activity can readily bring power station plumes down to ground level when solar insolation is highest during the middle of the day.

32. When peaks in sulphur dioxide have been observed at the London monitoring sites, they have usually been associated with the plumes from the Thameside power stations. Continuous sulphur dioxide monitoring has also been carried out in rural Staffordshire. Again peak concentrations were associated with the large coal-fired power stations in the vicinity of the monitoring site.

Small Combustion Plant

33. In addition to the large combustion plant, which produce by far the largest amounts of sulphur dioxide, but discharge through tall stacks so that ground level exceedence of health based guidelines are relatively rare, there are many thousands of smaller plant which are potentially associated with high ambient air concentrations. Plumes from these plant individually cover much smaller areas, and tend to lead to lower peak concentrations, but can nevertheless potentially cause or contribute to exceedences in local areas.

Diffuse Sources

34. Long term average concentrations are highest in areas where domestic coal burning is prevalent. The sites with the highest annual average data from the BUN and EC Directive network are all either in traditional coal mining areas or domestic solid fuel burning areas without natural gas (70 % of Northern Ireland homes use solid fuel for domestic heating).

35. The high annual averages at these sites reflect high background levels of sulphur dioxide due to diffuse sources such as high domestic and small boiler solid fuel use. The peaks due to larger plumes, when added to this background, tend to result in larger and more frequent exceedences of the short term standards as well. For instance, Belfast, which has local large combustion sources, high domestic solid fuel use, no natural gas, and unfavourable geography, regularly experiences concentrations of sulphur dioxide among the highest in the UK whatever averaging period is used for the data. This in turn means that exceedences of health based objectives are among the greatest there too.

36. Modelling of the releases from houses burning coal has shown that domestic emissions in even small communities are likely to result in exceedences of the EPAQS recommendation wherever a significant proportion of homes use coal for heating. For instance, in an area 250m square it is estimated that around 90 houses burning coal with a sulphur

content of 1.6% would be enough to enable the EPAQS recommendation to be exceeded. This corresponds to a housing density of 15 houses per hectare. Terraced housing typically has around 40 houses per hectare. For larger communities it is expected that a relatively smaller proportion of houses burning coal would be sufficient to cause exceedences of the recommendation.

37. The effect of vehicle emissions on sulphur dioxide levels can be seen by comparing the data from the Cromwell Road and Bridge Place sites in the AUN. The two sites are only around 2 km apart, the former being a kerbside site on a busy London trunk road and the latter an urban background site. Given that AUN sites are chosen so as not to be unduly influenced by local point sources, it is reasonable to assume that the major difference between the two will be due to the influence of traffic. The 1993-94 data for the two sites are shown in Figure II.10.4 in the form of graphs of hourly average concentrations. From this figure it can be seen that there is no significant difference in the time or size of the peak levels, suggesting a common remote source, possibly large combustion plant in the East Thames corridor. However the monthly arithmetic mean values show the kerbside site consistently higher by a few ppb (3-12).

Figure II.10.4:
The effect of traffic on urban sulphur dioxide levels

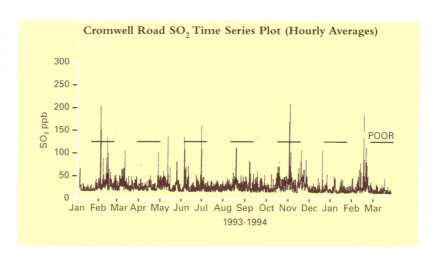

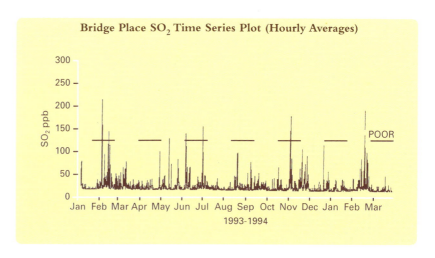

38. Thus it would appear that traffic does not significantly affect peak levels of sulphur dioxide, but does elevate the mean level near busy roads.

The Strategy

39. There are a number of policies in place dealing with individual sulphur dioxide emission source sectors, and these are discussed in turn below. Total national emissions of sulphur dioxide are the subject of the second Sulphur Protocol under the United Nations Economic Commission for Europe (UNECE) Convention on Long Range Transboundary Air Pollution (LRTAP). The UK signed this Protocol in Oslo in 1994, ratifying it in December 1996[3], thus committing the UK to reduce its total sulphur dioxide emissions by 50% by 2000, 70% by 2005, and 80% by 2010, on a 1980 base. Since the major source of sulphur dioxide emissions in the UK is the electricity supply industry and other large plant, the EC Large Combustion Plant Directive is an important instrument for delivering these commitments.

Large Combustion Plant

40. Sulphur dioxide emissions from large combustion plants are controlled by quotas laid out in the United Kingdom National Plan, established under the Environmental Protection Act 1990 (EPA 90). This is in response to the EC Large Combustion Plants Directive (88/609/EEC), which requires the United Kingdom to achieve a 60% reduction over 1980 levels from these sources by 2003.

41. This policy is unable in itself to prevent exceedences of the objective occurring as a result of large combustion plant plumes. This is because neither the National Plan, nor the UNECE targets, guarantee abatement of every plant, but instead constrain the annual emissions from a company or sector. This reflects the fact that both are aimed at reducing sulphur deposition as a result of long range transport, for which the total sulphur emitted over an annual period is the determinant of exposure, rather than peak localised air concentrations.

42. The United Kingdom National Plan has recently been amended to provide a series of aggregated annual national emission targets for sulphur dioxide from large combustion plant. In England, Scotland and Wales authorisations under the system of Integrated Pollution Control (IPC), established by EPA 90, are required from the relevant regulatory authority (the Environment Agency (EA) in England and Wales, and the Scottish Environment Protection Agency (SEPA) in Scotland) in order to operate the certain prescribed processes, including all those which fall within the National Plan. The authorisations contain conditions enforcing the terms of the National Plan, and may include allocations for permitted levels of sulphur dioxide emissions in tonnes in line with National Plan requirements. Northern Ireland plans to introduce similar measures to those in the rest of the UK over the next two years, but at present the terms of the National Plan are enforced by the Alkali Inspectorate.

43. Around 80% of current sulphur dioxide emissions are regulated in some form by EA/SEPA under IPC. Most of those emissions arise from combustion processes falling within the National Plan. EPA 90 obliges EA/SEPA to apply BATNEEC in deciding on authorisations on a site-by-site basis.

44. Against this background, and as part of their statutory responsibilities under EPA 90, HMIP, whose statutory duties were transferred to the EA, carried out a review in 1996 of IPC authorisations for power stations in England and Wales. Power stations were the first processes to be

[3]*Reducing National Emissions of Sulphur Dioxide: A Strategy for the UK*, DoE, 1996.

authorised under IPC, so this review is the first four yearly review of its kind, picking up a requirement set out in the original authorisations. In Scotland, the four yearly review of authorisations for the power stations is on a later timescale and is due to be completed by the end of 1997. In Northern Ireland, power stations will be authorised by the middle of 1998.

45. As power stations currently account for around 67% of national sulphur dioxide emissions, the revised authorisations, which set targets for future upgrading requirements, are of particular significance in determining likely future levels of emissions of sulphur dioxide. They will also have an important indirect effect in setting a precedent for the review of authorisations in other industrial combustion plant. The revised IPC authorisations oblige the generators in England and Wales to make significant reductions in the levels of their pollutant emissions. Overall limits are set for emissions of sulphur dioxide from the portfolio of stations operated by each generator for 1996, 1999, 2001 and 2005. In order to provide the generators with flexibility to switch generation between stations two limits are set for sulphur dioxide releases at each individual station. A series of "A" limits defines the maximum permitted emissions to protect the local environment. Those limits must not be exceeded at any time, but in aggregate amount to a much higher figure than the overall limit set for the station portfolio. Within the upper bounds of the "A" limits there are a series of "B" limits for each station which can be varied by the operator and notified to HMIP (now the Environment Agency) provided the sum of the "B" limits does not exceed the overall limit for portfolio emissions. The "B" limits serve principally to protect the environment from the cumulative effect of long range air transport of sulphur and take account of critical loads. From 2005, the maximum permitted level of emissions from all power stations in England and Wales will be 365 kilotonnes.

46. The Environment Agency has already been discussing with the power generators in England and Wales how exceedences of the Strategy objective by power stations may be avoided. Extensive modelling work undertaken to inform these discussions indicates that the extent of exceedence at a point arising from a particular station will depend on: the local meteorological conditions (broadly related to its geographical position); the operating pattern; and the proximity of other stations or large emissions. Where reductions in emissions are required, a number of different techniques are available, including reductions in load during high risk periods, fuel switching or the application of new abatement technologies. Following completion of the current modelling studies, an air quality protocol will be developed for trial at one or more stations before being more widely implemented. It is expected that by 2005, exceedences, if any, should be slight.

47. The Agency has also been modelling the emissions from oil refineries. This preliminary work suggests that, for the sites considered, breaches of the Strategy objective are likely to be relatively minor. However, combinations of plant in close proximity, larger refineries, or those with different release characteristics from those already studied, may cause exceedences.

48. There are a number of other factors already in play which are likely to have an increasing influence on future levels of sulphur dioxide emissions.

50. Sulphur dioxide emissions from large combustion plant in England

and Wales have been reducing rapidly in recent years. In 1995 emissions of sulphur dioxide from large combustion plant within the National Plan had fallen by 55% compared with 1980 levels. This reduction is in large part due to the investment in combined cycle gas turbine (CCGT) stations by the generators. A further substantial tranche of CCGT generation is currently under construction or firmly committed to come into operation before the year 2000 and will result in further substantial reductions in emissions.

51. Power is now on stream from the generating sets fitted with Flue Gas Desulphurisation (FGD) at two of the UK's largest power stations at Drax and Ratcliffe. Cleaner generation from these stations is making an increasing contribution to national power supplies with the completion of the installation of FGD in 1996. In Scotland, Scottish Power is considering plans to fit FGD to Longannet in the future using a seawater scrubbing technique.

52. In Northern Ireland, natural gas is coming to the Ballylumford Power Station to allow conversion from oil firing before the end of 1997, and this, combined with the possible decommissioning of coal fired units at Belfast West Power Station as they come out of contract, will have a major impact on sulphur dioxide emissions and hence an improvement in local air quality.

53. Therefore, current policies will lead to dramatic cuts in the total amount of sulphur dioxide emitted in the UK from large combustion plant and should largely remove exceedences of the objective from this sector, apart from under rare meteorological conditions. The first review of the strategy will monitor progress towards this goal, and consider whether or not further action is warranted. As the impact of large combustion plant on ambient air quality lessens through the application of the National Plan and BATNEEC under IPC, and the increased use of natural gas in the electricity supply industry, so the exceedences caused by smaller combustion plant and diffuse sources will be thrown into sharper relief.

Small Combustion Plant

54. A few of the remaining sources of sulphur dioxide emissions not covered by the National Plan fall under local authority air pollution control as Part B processes under EPA 90 (individual boilers with a capacity of between 20 and 50 MW).

55. In addition there are EC Directives (e.g. 93/12/EEC) relating to the sulphur content of certain liquid fuels which place mandatory limits on the sulphur content of industrial gas oil and diesel. These ban the marketing of gas oil with a sulphur content of greater than 0.2 %. Other than this, ground level concentrations of sulphur dioxide from combustion plant of less than 20 MW capacity are controlled only by guidance notes on stack height.

56. The first advice of this kind was issued in 1963. The third and current edition of the 1956 Clean Air Act Memorandum on Chimney Heights was published in 1981. Similar advice is contained in HMIP Technical Guidance Notes, e.g. Dispersion D1 (published in 1993). Although the guidance has been largely re-written over the years to extend its coverage to include low sulphur fuels, other pollutants, and new combustion plant technology, the basis of chimney height calculation for combustion plant using sulphur bearing fuels has not changed. The guidance is based upon the presumption that the critical requirement for chimney heights is to avoid short term acute pollution effects. The concentration of concern used for sulphur dioxide was 170 ppb, averaged over a time period of 15 - 30 minutes (the models used at the time cannot be more precise). The models apply under normal meteorological conditions (i.e. those weather

conditions expected around 98 % of the time in the UK). The stack height derived by the method also depends on assumptions on the background concentration in the area of the stack, and the predicted emission rate from the process.

57. Since the guidance also takes account of the effect of nearby stacks and local background concentrations, in theory exceedences of the WHO short-term standard of 175 ppb should be rare. Of course, the data discussed in preceding sections show that exceedences of this standard occur primarily in those parts of the country where domestic coal burning is prevalent. Indeed, more exceedences are likely to be found as the AUN is extended as planned to over double its current size.

58. These exceedences are likely to be caused by one or more of a number of reasons:

■ incorrect application of the guidance;

■ higher than expected local background;

■ combination of plumes from different directions on different days;

■ exceedence conditions associated with a limited wind arc.

59. On the other hand there are a number of considerations which will tend to make a particular stack less likely to exceed the guideline value. For instance, if the sulphur content of the fuel used is lower than that for which the stack was designed, then that boiler will be less likely to produce the expected number of exceedences. This is the case for many older installations as, in general, over the last few decades average sulphur contents have reduced. For instance, typical gas oil sulphur content in the seventies was around 0.8%, whereas it is now 0.18%. Ironically, this suggests that boilers with more modern stacks are more likely to cause exceedences than some older equipment.

60. A different picture emerges if the ground level concentrations from small combustion plant are compared with the standard of 100 ppb on a 15 minute basis. Exceedences up to 60 - 80% of the time are possible, depending on the stack height, heat release, size of plant, sulphur content of the fuel and any existing background concentration of sulphur dioxide. The largest possibility of an exceedence is from small plant burning black fuel, mainly boiler plant of around 1 MW capacity and below, with the probability of an exceedence reducing with increasing plant size. For plant around 10 MW capacity the possibility of exceedences reduces to around 10% of the time or less. The area around a single source within which exceedences may occur varies from fractions of a square kilometre for small plant around 1 MW with low uncorrected stack heights around 10 m or less, up to 100 km^2 or more for large plants with stacks of 100 m height order.

61. As above, the frequencies of exceedences at a particular site are diminished by the relatively low frequency with which a plume passes over the same area. This reduces the frequency of exceedences to the order of a few percent for small plant and less than 1% for larger plant. It should be noted that these are worst case estimates, based on pessimistic assumptions of load factors and fuel sulphur contents. Nevertheless, it seems clear that small combustion plant could at present contribute to significant numbers of exceedences of the objective.

62. In these circumstances, a revision of the chimney heights guidance is likely to be needed to ensure that new stacks do not cause exceedences of the objective, but current legislation may well require a review anyway. It should result in relatively small marginal costs, but is essential to the long term achievement of the objective. Upgrading of existing plant must depend on BATNEEC criteria, but there are a number of reasons why many existing plant may not pose a problem by 2005: existing commercial pressures to move to gas; the gradual introduction of fuels with lower sulphur contents than those for which the stack was designed - a process that has already been underway for many years; and uncertainties over the magnitude of the current problem caused by such plant.

Diffuse Sources

63. The use of coal in the domestic sector is controlled through Smoke Control legislation. It is likely that the progressive move away from coal and smokeless fuels (which generally have only 20% less sulphur than domestic coal) over the period covered by the Strategy will substantially remove the exceedences of the objective from these sources.

64. In Northern Ireland, Regulations are in preparation to make illegal the sale of unauthorised fuel in smoke control areas and to limit the sulphur content of all solid domestic fuel. The latter legislation is to prevent the supply of high sulphur petroleum coke for domestic use. This fuel has a sulphur content around three times higher than normal domestic coal. The comments on the consultation draft by the Chamber of Coal Traders noted that this fuel is also marketed in Yorkshire and the North East of England. It may thus be necessary to extend these controls to areas outside Northern Ireland.

65. A particular problem identified in the consultation draft was the provision of free coal to miners, contributing to the high concentrations found in current and former mining communities. However, it now appears that this may not be a significant problem by 2005 as there is a sharp trend downwards in the numbers of people taking such fuel, from 144,000 in 1994 through 118,000 now, to a projected 90,000 by 1998. This is due not only to the decline of the coal industry, and consequent changes in the average age of ex-miners, but also to many claimants preferring cash in lieu as they turn to gas fired boilers. It is thought that many claimants may accept buy-out offers over the next few years, reducing the numbers still further. Nevertheless, it is probable that concessionary fuel will still contribute to exceedences of the proposed Strategy objective in a few residual areas in 2005.

66. Vehicle emission limits primarily intended to reduce levels of particles and the regulated gaseous pollutants, will have the knock-on effect of decreasing sulphur dioxide emissions from transport. For instance, the sulphur content of diesel fuel is controlled under the same EC Directive that limits gas oil (93/12/EEC). Under this Directive, the sulphur content of diesel fuel was reduced from 0.2% to 0.05% in October 1996, leading to a consequent drop in sulphur dioxide emissions from diesel powered vehicles.

EC Developments

67. There are a number of prospective changes in European Community legislation which, if enacted in the period up to 2010, could also have an important influence on the future level of national sulphur dioxide emissions:

Chapter II.10: Sulphur Dioxide

EC Acidification Strategy

68. At the Environment Council in December 1995, the Commission presented a report on Community measures to counter acidification. They were asked to develop this further with the aim of bringing forward, by mid-1997, a coherent strategy for tackling acidification in the EC. Discussions are now underway within the EC on this strategy.

69. The intention is that the strategy should be effects-based, drawing upon the scientific research which has been carried out in the UNECE to underpin the second Sulphur Protocol and which is now being used in preparation for negotiations on a second Nitrogen Protocol. The Commission has set up a joint working group with UNECE to try to ensure co-ordination and consistency.

70. The strategy will provide an opportunity to define and clarify: the nature of the relationship between the EC and UNECE; the role of the EC in relation to acidification; the targets for emission reductions over the medium term; and the fundamental role which the analysis of costs and benefits should play in evaluating initiatives and action programmes.

Review of the Large Combustion Plant Directive

71. The terms of the original Large Combustion Plant Directive required the Commission to bring forward proposals for the revision of new plant emission limits by July 1995. They have accordingly been undertaking background research in preparation for a review of the Directive and they have been consulting member states and the interest groups.

72. It is understood that it may now be the Commission's intention that the proposals for review should go wider than the existing Directive and draw on the approach of the framework Directive on Integrated Pollution Prevention and Control (96/61/EEC) in seeking to regulate the environmental impacts of LCP. It is not yet clear on what timescale these proposals will come forward, nor whether they will relate to the upgrading of existing plant as well as to new plant standards.

Sulphur Content of Liquid Fuels Directive

73. Formal proposals have not yet been published by the Commission, but it is understood that they are unlikely to propose amending the limits set for the sulphur content of gas oil in the earlier Directive or to introduce new limits on the sulphur content of aviation or bunkering fuel. However, it is likely that the proposals will seek to set a limit for the sulphur content of heavy fuel oil. The draft proposals published in 1994 proposed a 1% by weight limit on the sulphur content of heavy fuel oil.

74. Heavy fuel oil is sold for use in the UK in a limited number of power stations, and to fire industrial boilers and furnaces. Although demand has been falling steadily in recent years and is projected to continue falling because of the competition from gas, the UK remains the second biggest user of heavy fuel oil in the European Union after Italy. The burning of heavy fuel oil, including refinery fuel oil, accounted for around 19% of UK emissions of sulphur dioxide in 1994.

Economic Instruments

75. There is a widespread commitment within the EC to the development of market based instruments where these can provide a more cost-effective alternative to the traditional regulatory command and control means of achieving environmental objectives. In preparing proposals for an EC acidification strategy, the Government will be encouraging the Commission to examine the use of economic instruments and market signals to promote structural change.

Conclusions   76. The Government has decided to accept the EPAQS recommendation of 100 ppb measured on a 15 minute averaging time basis as the standard for sulphur dioxide. The Government has also decided to adopt as a provisional objective the achievement of this standard at a 99.9th percentile by 2005, i.e. on all but 35 periods of 15 minutes, assuming complete data capture in a year

77. In view of the short averaging time of the objective for sulphur dioxide, the Government has decided that the objective should apply at any near- ground level non-occupational location outdoors where a person might reasonably be expected to be exposed over the averaging time of the objective.

78. The aim will therefore be to secure reductions sufficient to tackle residual background levels in domestic coal burning areas and to tackle localised plume grounding from combustion sources. To be sure of meeting the provisional objective, it will be necessary to reduce sulphur dioxide emissions in domestic coal burning areas. The Government will be considering the options for achieving this in the near future. It will also be vital to control the emissions of large combustion plant under IPC authorisations to levels that are consistent with the requirements of the EPAQS recommendation, using the concept of BATNEEC. To a certain extent the requirements of the Oslo Protocol and EPA 90 will encourage this regardless of ambient air quality objectives, but it is important to bear in mind that such objectives require that the rate of emission is controlled as well as the overall total emitted.

Annex 1: The Costs and Benefits of Air Quality Objectives

Background/ Context

1. In Chapter 4 it is stated that 'it is a fundamental principle of all Government policy that measures which incur a cost should achieve equivalent or greater benefits, and that the option taken could not be substituted by another which achieves the same benefit at less cost.' The text, however, realistically concedes that 'the principle is .....subject to limitations, not least the fact that benefits are often hard to define in monetary terms', but stresses that 'there should ...be an assessment of the merits of actions and the options for action in terms of costs and benefits, as far as it is possible and appropriate'.

2. The purpose of this Annex is to review progress in assessing the costs and the benefits of improvements in air quality, as an important underpinning of the National Air Quality Strategy (NAQS). The final section of this paper sets out a programme of studies which will inform the review of the Strategy in 1999.

The Costs of Meeting Air Quality Objectives

3. Relevant studies to develop appropriate cost estimates for air pollution abatement have been carried out by the UK Government, the European Commission and the United Nations Economic Commission for Europe (UNECE).

EC Work

4. A substantial programme of work on cost-effective road fuel and vehicle standards has been carried out under the European Commission's Auto Oil Programme. This represents an effective, collaborative venture between the Commission, and the European oil and vehicle industries. Key data produced under the Auto Oil Programme includes cost figures for the abatement of nitrogen oxide and particle emissions by 20-40% and 35-50% respectively. For example, the estimated increase in new vehicle costs would be around £210 for petrol cars and £295 for diesel cars (proposals for goods vehicles have not yet been finalised). Also, in terms of meeting more stringent motor fuel quality, costs per litre of fuel are quoted as 0.14p, and 0.59p per litre, for petrol and diesel respectively.

5. Such figures as those quoted above provide useful indications of the cost of the proposed EU directives on car and fuel standards, which are set to make a significant contribution towards the attainment of UK air quality standards for certain pollutants.

6. Information is available from other sources on the costs of converting diesel vehicles to alternative fuels, such as compressed natural gas (CNG) and liquefied petroleum gas (LPG), which can reduce emissions of nitrogen and particles by up to 70%, with an associated conversion cost ranging from a few hundred pounds to £20,000, covering the car to bus spectrum. In addition, retrofitting exhaust after treatment systems for heavy goods vehicles - such as those which the Government intends to incentivise via a reduction in vehicle excise duty - could reduce particulate emissions from these vehicles by up to 90% at a current cost of £3,000 to £6,000. The costs of both the conversion of vehicles to use alternative fuels and the retrofitting of particulate traps are likely to fall over time as a result of economies of scale.

UK Work

7. The Department of the Environment has undertaken a substantial study of the cost of abating air pollutants to inform both domestic and international policy, in order to ensure that policies are cost-effective. Recent studies include the following:

Sulphur

- a study of policy instruments for achieving compliance with national targets under the UNECE Second Sulphur Protocol [carried out by a joint team of consultants from the Centre for Social and Economic Research into the Global Environment (CSERGE), the Science Policy Research Unit at the University of Sussex (SPRU) and Economics for the Environment Consultancy (EFTEC)]; and a short follow-up study by SPRU on cost-effective abatement of sulphur, using a market-based approach, compared with a regulatory one (SPRU);

- a study of the costs of meeting standards for ambient SO_2 concentrations recommended by EPAQS and WHO, for 2005 (AEA Technology);

- a compliance cost assessment for UK Refineries of the proposed EC Sulphur Liquid Fuels Directive (Environmental Resources Management);

Volatile Organic Compounds(VOCs)

- a study of the cost of complying with the proposed EC Solvents Directive, and potential cost saving from adopting a market-based approach [Aspinwall/National Economic Research Associates (NERA)];

- a study of the cost of complying with the proposed Stage 2 Petrol Vapour Recovery Directive, compared with on-board vapour recovery for petrol vehicles (Chem Systems Ltd);

- an assessment of the potential for abatement and an estimate of the associated cost, of VOC emissions in Europe (AEA Technology) - to inform the preparatory work on the UNECE Second Nitrogen Protocol;

- The Air Pollution Abatement Review Group (APARG) Report on the abatement of VOCs from stationary sources (AEA Technology, Ch 8: Cost of abatement technology);

Nitrogen Dioxide

- A compliance cost assessment for the necessary measures to achieve the specific objective for nitrogen dioxide by 2005, as laid down in the Strategy (AEA Technology);

Particles - PM$_{10}$

- As above, a compliance cost assessment for the necessary measures to achieve the specific objective for particles by 2005 (AEA Technology);

and for the DTI:

- a study of the costs (and benefits) of the reduction of VOC emissions, from stationary sources (ERM);

8. The above studies have helped to inform the Department of the Environment's position in terms of the cost of current or anticipated policy initiatives on air quality. Annex 2 includes the key cost data that emerge from these studies.

9. The first study listed in Annex 2 - 'Sulphur Dioxide Ambient Air Quality Study' - was aimed at informing the Strategy target for sulphur dioxide ; and the 'Compliance Cost Assessment of the Proposed Air Quality Regulations for Nitrogen Dioxide and Fine Particles (PM$_{10}$) focused on measures required to meet the Strategy targets, as the title suggests.

However, most of the other studies covered in the annexes were not set up explicitly to carry out cost-benefit assessments of ambient air quality standards. A number, for example, were commissioned to inform the cost-benefit work in relation to transboundary air pollution, under the UNECE Convention on Long Range Transboundary Air Pollution; and a number were designed to inform policy on specific EC directives covering air pollution. Given the diversity of studies and their different contexts, the lack of uniformity in the presentation of results is to be expected: some studies report average costs, some marginal, some totals.

The Benefits of Meeting Air Quality Objectives

10. The valuation of the benefits of improved ambient air quality is generally associated with greater degrees of uncertainty than that associated with the valuation of costs. The robustness of benefit estimates is reliant not only on the reliability and accuracy of unit economic valuations, but also on the robustness of the underlying data on health and physical impacts.

11. However, the assessment of the benefits of improved air quality has attracted considerable attention among environmental economists, and substantial work has been produced both in the US and Europe. In informing the objectives laid down in the NAQS, it has been possible to draw on this external literature, as well as dedicated studies commissioned by the Department. Annex 2 documents the main, relevant studies on the benefits side for sulphur dioxide, nitrogen dioxide, ozone and fine particles. As in the case of work on the costs of abatement, studies commissioned for the UK Government (eg 'Acid Rain in Europe: Counting the Cost') and by the European Commission (eg 'Externe: Externalities of Energy', and cost-benefit work on the Air Quality Framework Directive) were drawn upon.

12. This is a constantly evolving area of work, and progress is being made in improving the underlying methodology, and in the precision of empirical estimates. The Government is encouraging and developing further work in this area. The Department of the Environment-led interdepartmental Group on Environmental Costs and Benefits (GECB) recently discussed the state of play in the academic literature of valuing the external costs of road transport, which include those associated with air pollution. The Group is also working to improve the robustness of estimates of the health benefits ensuing from improved air quality. Elsewhere, a major research study is also being carried out for HSE on the value of statistical life across different contexts.

13. At the European level, the Commission is funding additional work through the ExternE programme - the EC Study on the Externalities of Energy. The ExternE Programme uses a bottom-up methodology which was adopted initially for the core work on fuel cycles associated with electricity generation. The additional work aims to reduce some of the uncertainties associated with the original estimates, and to apply the methodology to other sectors, including the assessment of the external impacts of transport. Given the experience of the team working on this project, and the level of funding being provided, it is hoped that this work will take forward our understanding of the issues involved in benefit valuation.

Bringing the Cost and Benefit Sides together

14. Ideally, air quality objectives should be set on the basis of a comparison between the marginal costs and marginal benefits of abatement; for the most efficient outcome, pollution should be reduced up to the point at which the additional cost of abating the last unit is just equal to the additional benefit gained by that incremental change. In contrast, total costs (benefits) are the overall costs (benefits) associated with a given level of abatement; and average costs (benefits) are derived from the total cost (benefit) figure divided by the total number of units abated. Estimates of the marginal costs and benefits of small changes from the current position can be viewed with more confidence than estimates of large changes. However, in practice, it is often difficult to derive costs and benefits at the margin, since they will vary according to the starting position. Therefore, total and average costs are frequently used, as imperfect proxies, in guiding decision-makers on the appropriate level of abatement. This level of aggregation will disguise variation in costs and benefits of abatement by location and by time. Gaps in our estimates of marginal costs and benefits - notably for ozone and nitrogen - will need to be plugged in order to inform the 1999 Review.

15. The NAQS objectives were based on an assessment of the benefits - in keeping with the effects-based approach - and a consideration of the associated costs of compliance. It should be noted that due to the difficulties - discussed here and elsewhere - involved in monetising the benefits, a fully monetised cost-benefit analysis was not carried out.

Levels of Confidence in the Reported Estimates

16. Para 10 above made the point that often greater certainty can be attached to the cost estimates as compared with the monetary valuation of benefits. For the latter, it is important to recognise the scope for uncertainties to arise at each stage of the valuation process. The following steps are involved:

- modelling the impact of emissions on ambient concentrations of the relevant pollutant;

- establishing the epidemiological relationship between ambient concentrations and the physical impact on human health, buildings and materials, crops, forests and natural ecosystems - this is likely to involve transferring exposure-response relationships, established in other studies, often carried out in other countries;

- the definition of the stock at risk, which in the case of human health, for example, relates to the relevant population exposed;

- the unit valuation estimate to apply to the physical impact.

Given the uncertainties underlying each step, the final benefit estimates need to be viewed with some caution, and as indicative rather than definitive values. For example, very similar dose-response functions can be transformed into very different benefits estimates according to the valuations employed, a typical example being the values of a statistical life which different experts employ.

17. More primary work is undoubtedly needed to improve on the robustness of most of the benefit estimates, and to assess the robustness of each step in the estimation process. In Annex 2, the results from the studies already completed on both the costs and the benefits side are recorded, and the intention has been, where possible, to give an indication

of the range of values, not solely a point estimate, as the latter may be interpreted as implying more certainty than is warranted.

18. One benefits study for the Department of the Environment - 'Research into Damage Valuation Estimates For Nitrogen Based Pollutants' addresses the issue of uncertainty, by using meta analysis and Monte Carlo simulation techniques[1]. The results are then reported as ranges of values, with a 95% confidence interval.

19. As a final, general point on the material recorded in Annex 2, it is important to recognise the trade-off between conciseness and thorough coverage of riders and caveats; this Annex leans to conciseness.

Double Counting and Overstatement of Costs and Benefits

20. Double counting can be a problem associated with the measurement of both costs and benefits. For example, one issue which needs to be examined carefully is that of how to apportion costs between pollutants when a measure results in more than one pollutant being abated. For example, there would clearly be double-counting if the costs of certain measures applied to road vehicles were attributed to both NOx and particles, but this type of calculation has frequently been carried out, resulting in overestimation of the costs per unit of pollutant abated. Furthermore, where benefits occur in addition to those associated with better air quality, such as improved safety and reduced congestion, these should also be taken into account when allocating costs. It may indeed be the case that some policies, which result in reduced air pollutants as a secondary effect, are primarily motivated by other policy goals, in which case allocating such policy costs entirely to air pollution abatement would be inappropriate. This kind of simplistic approach is, however, not infrequently adopted, and policy-makers need to be alert to the consequent overstatement of costs per tonne of abated air pollutant.

21. A possible means of overcoming the double-counting problem in relation to costs may be to follow the approach used in the Auto Oil Programme of evaluating the costs of packages of measures, rather than contriving to apportion costs to specific pollutants. This approach has been adopted in carrying out the compliance cost assessments for NO_2 and PM_{10} - see reference in paras 4-6 above and in Annex 2.

22. Costs can also be biased upwards where the data collected has relied largely on the ex ante assessment provided by the industry/sector affected. Experience has shown that, in practice, ex post costs are often considerably lower than those estimated prior to compliance. This is particularly the case where a long lead time allows the industry affected to innovate, adapt and build the new requirements into its investment cycle.

23. In the case of benefits, where levels of individual pollutants tend to be associated with each other, either because they come from a common source or build up together in calm weather, analysis of any effects is made more difficult. For example, in a winter episode, it may not be possible to distinguish whether the health effects remaining after exclusion of other factors such as respiratory infections or cold weather are due to NO_x or particles. In such cases it would not be valid to apportion the health effects to both NO_x and particles and then add them up.

[1] Meta-analysis is a maximum likelihood technique which takes account of the different sample sizes and standard errors in the original studies, giving greater weight to those with bigger sample sizes and smaller standard error. Monte Carlo techniques involve the use of models in repeated experimentation and simulation in order to build up by the probability values of the different outcomes.

Concluding Comments and suggestions for future work

24. The aim of this paper has been to indicate the breadth of information on the costs and benefits of abating some of the major air pollutants covered in the consultation paper on the NAQS. It would be unrealistic to suggest that the information is perfect and complete, but the information contained in Annex 2 gives a flavour of the extent of relevant work already completed in this area, and points to gaps that need to be filled in informing the review of the Strategy in 1999.

25. Further work in this area should, therefore, be of a more dedicated type aimed at explicitly deriving comparable costs and benefits estimates for incremental improvements in ambient air quality. For the purposes of informing the review in 1999, a pollutant by pollutant approach is needed. Annex 2 aims to bring together the data on costs and benefits drawn from the current knowledge base. The next step should involve working up annualised costs and benefits of incremental improvements in ambient air quality, by pollutant, up to 2005. New data on emerging technologies and processes are likely to be an important element in informing the cost side, as is the development of a more refined time profile of the benefits accruing over time.

26. However, as raised in para 20, further consideration is also needed of the apportionment of costs between pollutants, the treatment of secondary benefits, and the possible interactions between pollutants. In preparing for the review of the Strategy, it should also be possible to draw directly on the major study currently being carried out for DG XI - 'Economic Evaluation of Air Quality Standards for SO_2, NO_2, fine particles and lead' - in support of the Air Quality Framework Directive.

27. An important step in advancing the benefits side is the recent decision by the Department of Health to form an expert group of health economists to advise on whether monetary valuation of the health effects of air pollution is appropriate, and, if so, what is the best approach to use in deriving estimates. The group will draw on the report from the sub-group of the Committee on the Medical Effects of Air Pollutants (COMEAP) which has a remit to quantify the various health effects expected from air pollution levels across the UK. This sub-group is due to report in June 1997; the health economists group would be expected to report in the summer of 1998.

Annex 2 – Summary of the Costs and Benefits of Abating Air Pollution as Covered by Recent Studies

The data below is drawn from studies on the costs/benefits of abating specified pollutants; these studies helped to inform the Strategy, but were not used to determine the objectives. Where possible, for comparability, the data has been converted into 1995 prices. Many of the benefit studies relate to large changes in emissions - with estimates often derived from total damage figures, not from considering small incremental changes.

POLLUTANT	£M pa	£/tonne abated (or per other unit specified)	Study	Comments
Sulphur Dioxide *Costs* i. EPAQS (100%) ii. EPAQS (99.9%) iii. EPAQS (99.8%) iv. WHO (100%) v. WHO (99.9%) vi. WHO(99.8%)	 76-166 65 57 37 9 7	213-441[1]	*1. Sulphur Dioxide Ambient Air Quality Study (AEA, September 1996), final report to DoE*	The study computed the annualised least cost of complying with the EPAQS standard for SO_2 in 2005, using an 8% discount rate. [1] represents the least cost per tonne abated from all relevant stationary sectors, using a levellised cost approach.
Benefits		1,329-2,372[2] (not directly related to ambient air quality standards, but to the damage, as a whole, from the emissions of tonnes of SO_2)	*2. Acid Rain in Europe: Counting the Cost (Helen ApSimon and David Pearce, Editors, 1995) report to the DoE, Earthscan publication early 1997.*	The study focussed on the benefits of abating transboundary air pollutants within Europe, and is based on benefit transfer of dose-response functions and unit valuation estimates from the established literature. The EXTERNE work is heavily drawn on. [2] represents the average benefit to the UK, for human health, buildings, crops and forests of abating one tonne of SO_2 in the UK. The figure is a snap-shot, once-off figure, no discounting has been applied.

POLLUTANT	£M pa	£/tonne abated (or per other unit specified)	Study	Comments
Ozone *Costs* of abating VOCs, as a precursor to ground level ozone	241.9 - 994.0[3]	353 -1451[3]	*3. APARG Report on the Abatement of Volatile Organic Compounds from Stationary Sources (AEA, July 1996)*	The report addresses the abatement of VOC emissions from stationary sources betweeen 1988 and 1999, the period covered by the UNECE VOC Protocol. [3] This represents the average cost per tonne abated associated with all measures to reduce or prevent emissions, in achieving a 35% reduction in emissions from stationary sources between 1988 and 1999. The costs are expressed in 1994 prices; capital costs are annualised over 5 years at a discount rate of 7.5% (except for Stage I and Stage II Controls where the discount rate is 6%)
		I. 660[4] ii. 280[5] iii 8,800[6]	*4. Costs and Benefits of the Reduction of VOC Emissions (ERM, February 1996)* report to the DTI	[4] represents the average cost per tonne abated, under the existing commitments (35% reduction); [5] represents the average cost per tonne, assuming the equivalent reduction (35%), but using the least cost approach. [6] represents the average cost per tonne of reducing emissions by 55% below 1988 levels - the 'maximum cut' scenario.
Benefits from reduction of ground level **ozone**	4,342 - 4,517[7] (a total damage estimate)		*5. Green Accounting Research Project, UK Case Study,(M. Holland et al, ETSU)*	[7] represents the total annual damage to the UK (human health, crops and materials) from existing ambient concentrations of ground level ozone.

Annex 2 – Summary of the Costs and Benefits of Abating Air Pollution as Covered by Recent Studies

POLLUTANT	£Mpa	£/tonne abated (or per other unit specified)	Study	Comments
Fine Particles (PM$_{10}$) *Costs*	[£1,049m[8] in total, not per annum - this includes the cost of abating 3,172 tonnes of PM$_{10}$ and 41,705 tonnes of NO$_x$ from mobile sources]		*6. A Compliance Cost Assessment of the Proposed Air Quality Regulations for Nitrogen Dioxide and Fine Particles (PM$_{10}$), (AEA), a report to the DoE*	[8] As covered in *para 21* of the main text, a package of measures has been costed **which includes the abatement of both fine particles and NO$_2$**. The figures quoted here represent the cost of transport measures covered in the Auto Oil Programme reports for Phase One, and relate to a movement from EURO II to EURO III Regulations. Both fuel modifications and vehicle technology changes applied to both petrol and diesel vehicles have been covered. A net present value for the capital and operating costs has been computed using a 7% discount rate.
Benefits	10,526 - 26,112[9] (this is a total damage estimate)		*7. Assessing the Health Costs of Particulate Air Pollution in the UK (Pearce and Crowards, in 'Energy Policy' Vol 24, No 7, 1996)*	[9]The paper aims to value the total mortality and morbidity health damage to the UK population from exposure to PM10 levels currently experienced in the UK. The study draws on the US epidemiological studies for the *mortality dose-response function* (ie work by Joel Schwartz) - no threshold and a linear relationship are assumed. The number of urban deaths per annum in the UK range from 3346 (low estimate) to 6952 (high estimate) . Using a *value of statistical life* (VOSL) of £1.5m (provisional on the work of the expert group of health economists) results in the range of mortality damage costs of between £7.33bn and £15.37bn (for urban and rural mortality together). The *morbidity dose response functions* and the *unit valuations* are drawn from the meta study by Rowe et al (1995), with the latter adjusted to reflect differences in income between the US and UK. Restricted activity days and chronic bronchitis dominate the total morbidity damage figures. Total morbidity figures for the the UK range between £3.20bn and 10.74bn, yielding total health costs of between £10.53 and £26.11 bn, all converted to 1995 prices.

POLLUTANT	£M pa	£/tonne abated (or per other unit specified)	Study	Comments
PM$_{10}$ *Benefits (cont)* **Nitrogen Dioxide**		(£6.25-18.75)[10]	8. *'The Valuation of Environmental Externalities' - (Robert Tinch, a report for the Department of Transport, 1996)*	[10] this reports on the amount individuals would be willing to pay, per person, per annum per μgm^{-3} to reduce the morbidity and mortality risk. The dose-response functions used were the same as those used in study 7 above, but the Department of Transport's official VOSL (£812, 000) was applied.
Costs	[£1,089m[11] in total, not per annum - this includes the cost of abating 41,705 tonnes of NO$_x$, and 3,172 tonnes of PM$_{10}$ from mobile sources and 260,000 tonnes of NO$_x$ from large combustion sources]		9. *A Compliance Cost Assessment of the Proposed Air Quality Regulations for Nitrogen Dioxide and Fine Particles (PM$_{10}$) (AEA),* forthcoming report to DoE.	[11] see footnote 8 above - the same points apply. The £1,089m includes the £1,049m for the transport measures, as covered above; plus £40m of measures applied to stationary sources (large combustion plant)
Benefits		£720-805[12]	10. *Acid Rain in Europe: Counting the Cost (Helen ApSimon and David Pearce, Editors, 1995)* report to the DoE, Earthscan publication early 1997	See general reference to the scope for this study under the benefits section for sulphur dioxide (above). [12] represents the average benefit to the UK, in terms of reduced damage to human health and buildings, from abating one tonne of nitrogen dioxide. No discounting has been applied.

Glossary

Acidification	The decrease in pH of surface waters and soils
AERONOX	European Research Programme on Emissions from Aircraft
AEROTRACE	European Research Programme on Emissions from Aircraft
ALSPAC	Avon Longitudinal Study of Pregnancy and Childhood
APHEA	Short term effects of air pollution on health. An EC funded research programme
AQMA	Air Quality Management Area
AUN	Automatic Urban Network
BATNEEC	Best Available Techniques Not Entailing Excessive Cost
BPEO	Best Practicable Environmental Option
BPM	Best Practicable Means
BUN	Basic Urban Network
CCGT	Combined Cycle Gas Turbine
CEC	Commission of the European Communities
CHP	Combined Heat and Power
CNG	Compressed Natural Gas
CO	Carbon Monoxide
Cold blast cupolas	A process used in metalworking
COMEAP	Committee on the Medical Effects of Air Pollutants, reporting to DH
COT	Committee on Toxicity of Chemicals in Food, Consumer Products and the Environment, reporting to DH
CRI	Chemical Releases Inventory
DH	Department of Health
DoE	Department of the Environment
DoT	Department of Transport
DTI	Department of Trade and Industry
EA	Environment Agency
EC	European Community
EMAS	Eco-management and audit scheme
EPA 90	Environmental Protection Act 1990
EPAQS	Expert Panel on Air Quality Standards, reporting to DoE
EPEFE	European Programme on Emissions, Fuels and Engine Technologies
Epidemiology	Statistical studies of health effects in populations
ESI	Electricity supply industry
ETSU	Energy Technology Support Unit
EU	European Union
Eutrophication	The increase of nutrients in ecosystems
FGD	Flue Gas Desulphurisation
GDP	Gross Domestic Product
Genotoxic Carcinogens	Substances which can cause cancers by attacking the genetic material
GLC	Greater London Council
HMIP	Her Majesty's Inspectorate of Pollution - now part of the Environment Agency
HVLP	High Volume Low Pressure
ICAO	International Civil Aviation Organisation
IPC	Integrated Pollution Control
IPPC	Integrated Pollution Prevention and Control
kt/a	Kilotonnes per annum
LAPC	Local Air Pollution Control
LOAELs	Lowest Observable Adverse Effect Levels

LPG	Liquefied Petroleum Gas
LRC	London Research Centre
LRTAP	UNECE Convention on Long-Range Transboundary Air Pollution
LSS	London Scientific Services
MARPOL	International Maritime Organisation Marine Pollution Convention
Mesoscale	Distance scale in meteorology of the order of a few hundred kilometres
NAQS	National Air Quality Strategy
NERC	National Environmental Research Council
NETCEN	National Environmental Technology Centre
ngm^{-3} or ng/m^3	nanograms per cubic metre (1000 ng = 1µg)
NO	Nitric Oxide
NO_2	Nitrogen Dioxide
NO_x	Oxides of Nitrogen, sum of NO and NO_2
NPPG	National Planning Policy Guidance (in Scotland)
NRTF	National Road Traffic Forecasts
Olefins	Organic compounds, found in petrol and oil
OS	Ordnance Survey
PAHs	Polycyclic Aromatic Hydrocarbons
Plumbosolvent	Can dissolve lead
Plume grounding	When exhaust fumes from a tall chimney are blown down to ground level
PM_{10}	Particulate Matter 10 microns (millionths of a metre) or less in diameter
POCP	Photochemical Ozone Creation Potential
POLINAT	European Research Programme on Emissions from Aircraft
PORG	Photochemical Oxidants Review Group, reporting to DoE
ppb	parts per billion (parts per 1000 million)
PPGs	Planning Policy Guidance Notes
ppm	parts per million
PVR	Petrol Vapour Recovery
QUARG	Quality of Urban Air Review Group, reporting to DoE
'Real-time' monitors	Automatic monitors which give continuous updates, as opposed to diffusion tubes which provide figures averaged over longer time-scales
SACTRA	Standing Advisory Committee on Trunk Road Assessment, reporting to Department of Transport
SEPA	Scottish Environment Protection Agency
SO	Scottish Office
SO_2	Sulphur Dioxide
SPRU	Science Policy Research Unit
Synergistic	The whole is greater than the sum of the parts
Toxicology	Study of compounds that are poisonous to humans
μgm^{-3} or $\mu g/m^3$	Micrograms per cubic metre
µm	1 µm = 1 micron = 1 millionth of a metre
UKMO	United Kingdom Meteorological Office
UNECE	United Nations Economic Commission for Europe
VED	Vehicle Excise Duty
VOCs	Volatile Organic Compounds
WHO	World Health Organisation
WO	Welsh Office

Printed in the UK for The Stationery Office Limited on behalf of the
Controller of Her Majesty's Stationery Office
Dd 5065700   3/97   3401/A   76368